Introduction to Atomic Physics

Introduction to Atomic Physics

Harley Reynolds

NY RESEARCH PRESS

New York

Published by NY Research Press
118-35 Queens Blvd., Suite 400,
Forest Hills, NY 11375, USA
www.nyresearchpress.com

Introduction to Atomic Physics
Harley Reynolds

International Standard Book Number: 978-1-63238-892-6 (Hardback)

Cataloging-in-Publication Data

Introduction to atomic physics / Harley Reynolds.
    p. cm.
Includes bibliographical references and index.
ISBN 978-1-63238-892-6
1. Nuclear physics. 2. Physics. 3. Nuclear energy. I. Reynolds, Harley.
QC776 .I58 2022
539.7--dc23

TABLE OF CONTENTS

This book has been written, keeping in view that students want more practical information. Thus, my aim has been to make it as comprehensive as possible for the readers. I would like to extend my thanks to my family and co-workers for their knowledge, support and encouragement all along.

Atomic physics deals with the study of atoms as an isolated system of electrons and an atomic nucleus. Atomic models in this discipline consist of a single nucleus that can be surrounded by one or more bound electrons. These electrons exist in a notional shell around the nucleus, and are considered to be in bound state. Atomic physics deals with the study of processes such as ionization and excitation by photons and collisions with atomic particles. The process of ionization involves an atom or molecule obtaining negative or positive charge by gaining or losing electrons. Atomic physics is closely related to the fields of molecular and optical physics. This book explores all the important aspects of atomic physics in the present day scenario. The topics covered herein deal with the core subjects of this field. The coherent flow of topics, student-friendly language and extensive use of examples make this book an invaluable source of knowledge.

A brief description of the chapters is provided below for further understanding:

Chapter – Atoms and Atomic Structure

The smallest unit of any molecule or ordinary matter is termed as an atom. It comprises of subatomic particles such as electrons, protons and neutrons. Atomic number, atomic mass, Brownian motion, etc. are some of its aspects. This is an introductory chapter which will briefly introduce all these significant aspects of atoms and atomic structure.

Chapter – Fundamentals of Atomic Physics

The scientific study of atoms and their structures, arrangements, their energy states and interactions with other particles is termed as atomic physics. Some of its components include atomic units, atomic models, isotopes, isobars, atomic spectroscopy etc. All the diverse components of atomic physics have been carefully analysed in this chapter.

Chapter – Interactions of Atoms

Depending upon the frequency and wavelength, an atom shows several types of behavioral aspects when exposed to different radiations. Some of the phenomena which are considered during this exposure are ground state of atom, excited state of atom, emission and absorption spectra. These phenomena elaborated in this chapter will help in gaining a better perspective about the subject.

Chapter – Bohr Model of the Atom

Bohr model represents the structure of atom and its constituents. It depicts nucleus at the center and electrons revolve around it. This chapter discusses in detail the theories and methodologies related to Bohr model including Bohr model of atom and hydrogen, hydrogen spectrum, Franck-Hertz experiment, etc.

Chapter – Quantum Mechanical Model of the Atom

Quantum mechanical model is the model of an atom which is based on mathematics and quantum theory. It is primarily used to explain the observations made on complex atoms. This chapter closely examines the key concepts of quantum mechanical model of atom to provide an extensive understanding of the subject.

Harley Reynolds

Atoms and Atomic Structure

The smallest unit of any molecule or ordinary matter is termed as an atom. It comprises of sub-atomic particles such as electrons, protons and neutrons. Atomic number, atomic mass, Brownian motion, etc. are some of its aspects. This is an introductory chapter which will briefly introduce all these significant aspects of atoms and atomic structure.

ATOMS

Atom is the smallest unit into which matter can be divided without the release of electrically charged particles. It also is the smallest unit of matter that has the characteristic properties of a chemical element. As such, the atom is the basic building block of chemistry.

Most of the atom is empty space. The rest consists of a positively charged nucleus of protons and neutrons surrounded by a cloud of negatively charged electrons. The nucleus is small and dense compared with the electrons, which are the lightest charged particles in nature. Electrons are attracted to any positive charge by their electric force; in an atom, electric forces bind the electrons to the nucleus.

Because of the nature of quantum mechanics, no single image has been entirely satisfactory at visualizing the atom's various characteristics, which thus forces physicists to use complementary pictures of the atom to explain different properties. In some respects, the electrons in an atom behave like particles orbiting the nucleus. In others, the electrons behave like waves frozen in position around the nucleus. Such wave patterns, called orbitals, describe the distribution of individual electrons. The behaviour of an atom is strongly influenced by these orbital properties, and its chemical properties are determined by orbital groupings known as shells.

Atomic Model

Most matter consists of an agglomeration of molecules, which can be separated relatively easily. Molecules, in turn, are composed of atoms joined by chemical bonds that are more difficult to break. Each individual atom consists of smaller particles—namely, electrons and nuclei. These particles are electrically charged, and the electric forces on the charge are responsible for holding the atom together. Attempts to separate these smaller constituent particles require ever-increasing amounts of energy and result in the creation of new subatomic particles, many of which are charged.

An atom consists largely of empty space. The nucleus is the positively charged centre of an atom and contains most of its mass. It is composed of protons, which have a positive charge, and neutrons, which have no charge. Protons, neutrons, and the electrons surrounding them are long-lived particles present in all ordinary, naturally occurring atoms. Other subatomic particles may be found in association with these three types of particles. They can be created only with the addition of enormous amounts of energy, however, and are very short-lived.

All atoms are roughly the same size, whether they have 3 or 90 electrons. Approximately 50 million atoms of solid matter lined up in a row would measure 1 cm (0.4 inch). A convenient unit of length for measuring atomic sizes is the angstrom $(\mathring{A})$, defined as 10^{-10} metre. The radius of an atom measures $1-2$ $\mathring{A}$. Compared with the overall size of the atom, the nucleus is even more minute. It is in the same proportion to the atom as a marble is to a football field. In volume the nucleus takes up only 10^{-14} metres of the space in the atom—i.e., 1 part in 100,000. A convenient unit of length for measuring nuclear sizes is the femtometre (fm), which equals 10^{-15} metre. The diameter of a nucleus depends on the number of particles it contains and ranges from about 4 fm for a light nucleus such as carbon to 15 fm for a heavy nucleus such as lead. In spite of the small size of the nucleus, virtually all the mass of the atom is concentrated there. The protons are massive, positively charged particles, whereas the neutrons have no charge and are slightly more massive than the protons. The fact that nuclei can have anywhere from 1 to nearly 300 protons and neutrons accounts for their wide variation in mass. The lightest nucleus, that of hydrogen, is 1,836 times more massive than an electron, while heavy nuclei are nearly 500,000 times more massive.

ATOMIC NUMBER

Atomic number is the number of protons in the nucleus of an element. It defines the position of the element in the periodic table. Atomic weight, which is a another number that appears next to the element's symbol, is an average of the atomic masses of all the isotopes of that element.

The number of protons in the nucleus determines the atomic number of an element. It's different from atomic mass or atomic weight, which take into account the presence of neutrons. Every atom of a given element always has the same atomic number, but atomic mass can vary according to the number of neutrons in the nucleus.

The Periodic Table

The periodic table is a chart that lists all the elements in order according to increasing atomic number. Scientists know of 118 elements. Number 118, oganesson (Og), which is an artificially produced radioactive element, was added in 2015. Oganesson has the highest atomic number because it has the highest number of protons in its nucleus. Hydrogen (H), on the other hand, has only one proton in its nucleus, so its atomic number is 1, and it appears at the beginning of the periodic table. The atomic number of each element, which is the number of protons in its nucleus, appears next to its symbol in the table. If the atomic number wasn't there, you could still tell how many protons were in the nucleus of a given element by counting the number of places between that element and hydrogen.

Atomic Number is not Atomic Mass or Atomic Weight

If one looks up an element in the periodic table, one will see another number next to its atomic number. This is the atomic weight of the element, and it's usually twice the atomic number or more. Atomic weight isn't the same as atomic mass.

The atomic mass of an atom is the mass of all the protons and neutrons in the nucleus. Electrons have such small masses compared to nucleons that they are considered negligible. Atomic mass is expressed in atomic mass units (amu) for a single atom and in grams per mole for macroscopic quantities. A mole is quantified as Avogadro's number $\left(6.02 \times 10^{23}\right)$ of atoms.

An atom of a given element always has the same number of protons. If it had a different number, it would be a different element. However, atoms of the same element can have different numbers of neutrons. Each version is called an isotope of that element, and each isotope has a different atomic mass. The atomic mass listed in the periodic table is an average of the atomic masses of all the naturally occurring isotopes of that element. This average is the atomic weight for that element.

ATOMIC MASS

Mass is a basic physical property of matter. The mass of an atom or a molecule is referred to as the atomic mass. The atomic mass is used to find the average mass of elements and molecules and to solve stoichiometry problems.

In chemistry, there are many different concepts of mass. It is often assumed that atomic mass is the mass of an atom indicated in unified atomic mass units (u).

The name "atomic mass" is used for historical reasons, and originates from the fact that chemistry was the first science to investigate the same physical objects on macroscopic and microscopic levels. In addition, the situation is rendered more complicated by the isotopic distribution. On the macroscopic level, most mass measurements of pure substances refer to a mixture of isotopes. This means that from a physical stand point, these mixtures are not pure. For example, the macroscopic mass of oxygen $\left(O_2\right)$ does not correspond to the microscopic mass of O_2. The former usually implies a certain isotopic distribution, whereas the latter usually refers to the most common isotope $\left({}^{16}O_2\right)$. Note that the former is now often referred to as the "molecular weight" or "atomic weight".

Mass Concepts in Chemistry

Name in chemistry	Physical meaning	Symbol	Units
Atomic mass	Mass on microscopic scale	m, m_a	Da, u, kg, g
Molecular mass	Mass of a molecule	m	Da, u, kg, g
Isotopic mass	Mass of a specific isotope		Da, u, kg, g
Mass of entity	Mass of a chemical formula	m, m_f	Da, u, kg, g
Average mass	Average mass of a isotopic distribution	m	Da, u, kg, g

Molar mass	Average mass per mol	$M = m/n$	kg/mol or g/mol
Atomic weight	Average mass of an element	$A_r = m \,/\, m_u$	unitless
Molecular weight	Average mass of a molecule	$M_r = m \,/\, m_u$	unitless
Relative atomic mass	Ratio of mass m and and the atomic mass constant m_u	$A_r = m \,/\, m_u$	unitless
Atomic mass constant	$m_u = m\left(^{12}C\right)/12$	$m_u = 1\,Da = 1\,u$	Da, u, kg, g
Relative molecular mass	Ratio of mass m of a molecule and and the atomic mass constant m_u	$M_r = m \,/\, m_u$	unitless
Relative molar mass	------	?	?
Mass number	Nucleon number	A	nucleons, or unitless
Integer mass	Nucleon number * Da	m	Da, u
Nominal mass	Integer mass of molecule consisting of most abundant isotopes	m	Da, u
Exact mass	Mass of molecule calculated from the mass of its isotopes (in contrast of measured ba a mass spectrometer)	------	Da, u, kg, g
Accurate mass	Mass (not normal mass)	------	Da, u, kg, g

Average Mass

Isotopes are atoms with the same atomic number, but different mass numbers. A different mass size is due to the difference in the number of neutrons that an atom contains. Although mass numbers are whole numbers, the actual masses of individual atoms are never whole numbers (except for carbon-12, by definition). This explains how lithium can have an atomic mass of 6.941 Da. The atomic masses on the periodic table take these isotopes into account, weighing them based on their abundance in nature; more weight is given to the isotopes that occur most frequently in nature. Average mass of the element E is defined as:

$$m(E) = \sum_{n=1} m(I_n) \times p(I_n)$$

where $\sum$ represents a n-times summation over all isotopes I_n of element E, and p(I) represents the relative abundance of the isotope I.

Example:

Find the average atomic mass of boron using the table below:

Table: Mass and Abundance of Boron Isotopes.

n	isotope I_n	mass m (Da)	isotopic abundance p
1	10_B	10.013	0.199
2	11_B	11.009	0.801

Solution:

The average mass of Boron is:

$$m(B) = (10.013\,Da) + (11.009\,Da)(0.801) = 1.99\,Da + 8.82\,Da = 10.81\,Da$$

Relative Mass

Traditionally it was common practice in chemistry to avoid using any units when indicating atomic masses (e.g. masses on microscopic scale). Even today, it is common to hear a chemist say, "^{12}C has exactly mass 12". However, because mass is not a dimensionless quantity, it is clear that a mass indication needs a unit. Chemists have tried to rationalize the omission of a unit; the result is the concept of relative mass, which strictly speaking is not even a mass but a ratio of two masses. Rather than using a unit, these chemists claim to indicate the ratio of the mass they want to indicate and the atomic mass constant mu which is defined analogous to the unit they want to avoid. Hence the relative atomic mass of the mass m is defined as:

$$A_r = \frac{m}{m_u}$$

The quantity is now dimensionless. As this unit is confusing and against the standards of modern metrology, the use of relative mass is discouraged.

Molecular Weight, Atomic Weight, Weight vs. Mass

Until recently, the concept of mass was not clearly distinguished from the concept of weight. In colloquial language this is still the case. Many people indicate their "weight" when they actually mean their mass. Mass is a fundamental property of objects, whereas weight is a force. Weight is the force F exerted on a mass m by a gravitational field. The exact definition of the weight is controversial. The weight of a person is different on ground than on a plane. Strictly speaking, weight even changes with location on earth.

When discussing atoms and molecules, the mass of a molecule is often referred to as the "molecular weight". There is no univerally-accepted definition of this term; however, mosts chemists agree that it means an average mass, and many consider it dimensionless. This would make "molecular weight" a synonym to "average relative mass".

Integer Mass

Because the proton and the neutron have similar mass, and the electron has a very small mass compared to the former, most molecules have a mass that is close to an integer value when measured in daltons. Therefore it is quite common to only indicate the integer mass of molecules. Integer mass is only meaningful when using dalton (or u) units.

Accurate Mass

Many mass spectrometers can determine the mass of a molecule with accuracy exceeding that of the integer mass. This measurement is therefore called the accurate mass of the molecule. Isotopes (and hence molecules) have atomic masses that are not integer masses due to a mass defect caused by binding energy in the nucleus.

Units

The atomic mass is usually measured in the units unified atomic mass unit (u), or dalton (Da).

Both units are derived from the carbon-12 isotope, as 12 u is the exact atomic mass of that isotope. So 1 u is 1/12 of the mass of a carbon-12 isotope:

$$1\,u \ = \ 1\,Da \ = \ m\left(^{12}C\right)/12$$

The first scientists to measure atomic mass were John Dalton (between 1803 and 1805) and Jons Jacoband Berzelius (between 1808 and 1826). Early atomic mass theory was proposed by the English chemist William Prout in a series of published papers in 1815 and 1816. Known was Prout's Law, Prout suggested that the known elements had atomic weights that were whole number multiples of the atomic mass of hydrogen. Berzelius demonstrated that this is not always the case by showing that chlorine (Cl) has a mass of 35.45, which is not a whole number multiple of hydrogen's mass.

Some chemists use the atomic mass unit (amu). The amu was defined differently by physicists and by chemists:

- Physics: $1\,amu \ = \ m\left(^{16}O\right)/16$

- Chemistry: $1\,amu \ = \ m\left(O\right)/16$

Chemists used oxygen in the naturally occurring isotopic distribution as the reference. Because the isotopic distribution in nature can change, this definition is a moving target. Therefore, both communities agreed to the compromise of using $m\left(^{12}C\right)/12$ as the new unit, naming it the "unified atomic mass unit" (u). Hence, the amu is no longer in use; those who still use it do so with the definition of the u in mind. For this reason, the dalton (Da) is increasingly recommended as the accurate mass unit.

Neither u nor Da are SI units, but both are recognized by the SI.

Molar Mass

The molar mass is the mass of one mole of substance, whether the substance is an element or a compound. A mole of substance is equal to Avogadro's number $\left(6.023\times10^{23}\right)$ of that substance. The molar mass has units of g/mol or kg/mol. When using the unit g/mol, the numerical value of the molar mass of a molecule is the same as its average mass in daltons:

- Average mass of C: $12.011\,Da$

- Molar mass of C: $12.011\,g\,/\,mol$

This allows for a smooth transition from the microscopic world, where mass is measured in daltons, to the macroscopic world, where mass is measured in kilograms.

Example:

What is the molar mass of phenol, C_6H_5OH ?

Average mass $m = 6 \times 12.011\,\text{Da} + 6 \times 1.008\,\text{Da} + 1 \times 15.999\,\text{Da} = 94.113\,\text{Da}$

Molar mass $= \ 94.113\,g\,/\,mol \ = \ 0.094113\,kg\,/\,mo$

Measuring Masses in the Atomic Scale

Masses of atoms and molecules are measured by mass spectrometry. Mass spectrometry is a technique that measures the mass-to-charge ratio (m/q) of ions. It requires that all molecules and atoms to be measured be ionized. The ions are then separated in a mass analyzer according to their mass-to-charge ratio. The charge of the measured ion can then be determined, because it is a multiple of the elementary charge. The the ion's mass can be deduced. The average masses indicated in the periodic table are then calculated using the isotopic abundances.

The masses of all isotopes have been measured with very high accuracy. Therefore, it is much simpler and more accurate to calculate the mass of a molecule of interest as a sum of its isotopes than measuring it with a commercial mass spectrometer.

Note that the same is not true on the nucleon scale. The mass of an isotope cannot be calculated accurately as the sum of its particles: this would ignore the mass defect caused by the binding energy of the nucleons, which is significant.

Table: Mass of Three sub-atomic Particles.

Particle	SI (kg)	Atomic (Da)	Mass Number A
Proton	1.6726×10^{-27}	1.0073	1
Neutron	1.6749×10^{-27}	1.0087	1
Electron	9.1094×10^{-31}	0.00054858	0

As shown in table, the mass of an electron is relatively small; it contributes less than 1/1000 to the overall mass of the atom.

The atomic mass found on the Periodic Table (below the element's name) is the average atomic mass. For example, for Lithium:

```
  3
 Li
      ↙
6.941
```

The red arrow indicates the atomic mass of lithium. As shown in table above and mathematically explained below, the masses of a protons and neutrons are about 1u. This, however, does not explain why lithium has an atomic mass of 6.941 Da where 6 Da is expected. This is true for all elements on the periodic table. The atomic mass for lithium is actually the average atomic mass of its isotopes.

One particularly useful way of writing an isotope is as follows:

$$_A^M E$$

M = Atomic Mass (Neutrons + Protons)

A = Atomic Number (Protons)

E = Element

Applications

Applications Include:

- Average Molecular Mass,

- Stoichiometry.

Note: One particularly important relationship is illustrated by the fact that an atomic mass unit is equal to 1.66×10^{-24} g. This is the reciprocal of Avogadro's constant, and it is no coincidence:

$$\frac{\text{Atomic Mass (g)}}{1\text{g}} \times \frac{1\text{mol}}{6.022 \times 10^{23}} = \frac{\text{Mass (g)}}{1\text{atom}}$$

Because a mol can also be expressed as gram × atoms,

$$1\ u = \frac{M_u (molar\ mass\ unit)}{(Avogadro's\ Number)} = 1 \frac{g}{mol\ N_A}$$

1u = Mu(molar mass unit)/N_A(Avogadro's Number)=1g/mol/N_A.

N_A known as Avogadro's number (Avogadro's constant) is equal to 6.023×10^{23} atoms.

Atomic mass is particularly important when dealing with stoichiometry.

ATOMIC NUCLEUS

The atomic nucleus is the small, dense region consisting of protons and neutrons at the center of an atom, discovered in 1911 by Ernest Rutherford based on the 1909 Geiger–Marsden gold foil experiment. After the discovery of the neutron in 1932, models for a nucleus composed of protons and neutrons were quickly developed by Dmitri Ivanenko and Werner Heisenberg. An atom is composed of a positively-charged nucleus, with a cloud of negatively-charged electrons surrounding it, bound together by electrostatic force. Almost all of the mass of an atom is located in the nucleus, with a very small contribution from the electron cloud. Protons and neutrons are bound together to form a nucleus by the nuclear force.

The diameter of the nucleus is in the range of 1.7566 fm $\left(1.7566 \times 10^{-15} m\right)$ for hydrogen (the diameter of a single proton) to about 11.7142 fm for the heaviest atom uranium. These dimensions are much smaller than the diameter of the atom itself (nucleus + electron cloud), by a factor of about 26,634 (uranium atomic radius is about 156 pm $\left(156 \times 10^{-12} m\right)$) to about 60,250 (hydrogen atomic radius is about 52.92 pm).

The branch of physics concerned with the study and understanding of the atomic nucleus, including its composition and the forces which bind it together, is called nuclear physics.

Nuclear Makeup

The nucleus of an atom consists of neutrons and protons, which in turn are the manifestation of more elementary particles, called quarks, that are held in association by the nuclear strong force in certain stable combinations of hadrons, called baryons. The nuclear strong force extends far enough from each baryon so as to bind the neutrons and protons together against the repulsive electrical force between the positively charged protons. The nuclear strong force has a very short range, and essentially drops to zero just beyond the edge of the nucleus. The collective action of the positively charged nucleus is to hold the electrically negative charged electrons in their orbits about the nucleus. The collection of negatively charged electrons orbiting the nucleus display an affinity for certain configurations and numbers of electrons that make their orbits stable. Which chemical element an atom represents is determined by the number of protons in the nucleus; the neutral atom will have an equal number of electrons orbiting that nucleus. Individual chemical elements can create more stable electron configurations by combining to share their electrons. It is that sharing of electrons to create stable electronic orbits about the nucleus that appears to us as the chemistry of our macro world.

Protons define the entire charge of a nucleus, and hence its chemical identity. Neutrons are electrically neutral, but contribute to the mass of a nucleus to nearly the same extent as the protons. Neutrons can explain the phenomenon of isotopes (same atomic number with different atomic mass.) The main role of neutrons is to reduce electrostatic repulsion inside the nucleus.

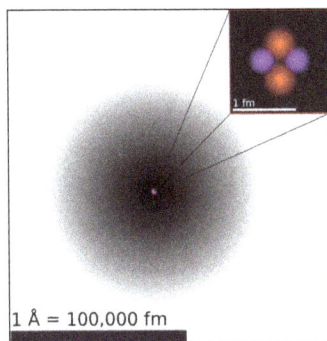

A figurative depiction of the helium-4 atom with the electron cloud in shades of gray. In the nucleus, the two protons and two neutrons are depicted in red and blue. This depiction shows the particles as separate, whereas in an actual helium atom, the protons are superimposed in space and most likely found at the very center of the nucleus, and the same is true of the two neutrons. Thus, all four particles are most likely found in exactly the same space, at the central point. Classical images of separate particles fail to model known charge distributions in very small nuclei. A more accurate image is that the spatial distribution of nucleons in a helium nucleus is much closer to the helium electron cloud shown here, although on a far smaller scale, than to the fanciful nucleus image.

Composition and Shape

Protons and neutrons are fermions, with different values of the strong isospin quantum number, so two protons and two neutrons can share the same space wave function since they are not identical quantum entities. They are sometimes viewed as two different quantum states of the same particle, the nucleon. Two fermions, such as two protons, or two neutrons, or a proton + neutron (the deuteron) can exhibit bosonic behavior when they become loosely bound in pairs, which have integer spin.

In the rare case of a hypernucleus, a third baryon called a hyperon, containing one or more strange quarks and other unusual quark(s), can also share the wave function. However, this type of nucleus is extremely unstable and not found on Earth except in high energy physics experiments.

The neutron has a positively charged core of radius $\approx$ 0.3 fm surrounded by a compensating negative charge of radius between 0.3 fm and 2 fm. The proton has an approximately exponentially decaying positive charge distribution with a mean square radius of about 0.8 fm.

Nuclei can be spherical, rugby ball-shaped (prolate deformation), discus-shaped (oblate deformation), triaxial (a combination of oblate and prolate deformation) or pear-shaped.

Forces

Nuclei are bound together by the residual strong force (nuclear force). The residual strong force is a minor residuum of the strong interaction which binds quarks together to form protons and neutrons. This force is much weaker between neutrons and protons because it is mostly neutralized within them, in the same way that electromagnetic forces between neutral atoms (such as van der Waals forces that act between two inert gas atoms) are much weaker than the electromagnetic forces that hold the parts of the atoms together internally (for example, the forces that hold the electrons in an inert gas atom bound to its nucleus).

The nuclear force is highly attractive at the distance of typical nucleon separation, and this overwhelms the repulsion between protons due to the electromagnetic force, thus allowing nuclei to exist. However, the residual strong force has a limited range because it decays quickly with distance thus only nuclei smaller than a certain size can be completely stable. The largest known completely stable nucleus (i.e. stable to alpha, beta, and gamma decay) is lead-208 which contains a total of 208 nucleons (126 neutrons and 82 protons). Nuclei larger than this maximum are unstable and tend to be increasingly short-lived with larger numbers of nucleons. However, bismuth-209 is also stable to beta decay and has the longest half-life to alpha decay of any known isotope, estimated at a billion times longer than the age of the universe.

The residual strong force is effective over a very short range (usually only a few femtometres (fm); roughly one or two nucleon diameters) and causes an attraction between any pair of nucleons. For example, between protons and neutrons to form [NP] deuteron, and also between protons and protons, and neutrons and neutrons.

Halo Nuclei and Nuclear Force Range Limits

The effective absolute limit of the range of the nuclear force (also known as residual strong force) is represented by halo nuclei such as lithium-11 or boron-14, in which dineutrons, or other collections of neutrons, orbit at distances of about 10 fm (roughly similar to the 8 fm radius of the nucleus of uranium-238). These nuclei are not maximally dense. Halo nuclei form at the extreme edges of the chart of the nuclides—the neutron drip line and proton drip line—and are all unstable with short half-lives, measured in milliseconds; for example, lithium-11 has a half-life of 8.8 ms.

Halos in effect represent an excited state with nucleons in an outer quantum shell which has unfilled energy levels "below" it (both in terms of radius and energy). The halo may be made of either neutrons [NN, NNN] or protons [PP, PPP]. Nuclei which have a single neutron halo include ^{11}Be and ^{19}C. A two-neutron halo is exhibited by $^{6}He,$ $^{11}Li,$ $^{17}B,$ ^{19}B and ^{22}C. Two-neutron halo nuclei break into three fragments, never two, and are called Borromean nuclei because of this behavior (referring to a system of three interlocked rings in which breaking any ring frees both of the others). ^{8}He and ^{14}Be both exhibit a four-neutron halo. Nuclei which have a proton halo include 8B and ^{26}P. A two-proton halo is exhibited by ^{17}Ne and ^{27}S. Proton halos are expected to be more rare and unstable than the neutron examples, because of the repulsive electromagnetic forces of the excess proton(s).

Nuclear Models

Although the standard model of physics is widely believed to completely describe the composition and behavior of the nucleus, generating predictions from theory is much more difficult than for most other areas of particle physics. This is due to two reasons:

- In principle, the physics within a nucleus can be derived entirely from quantum chromodynamics (QCD). In practice however, current computational and mathematical approaches for solving QCD in low-energy systems such as the nuclei are extremely limited. This is due to the phase transition that occurs between high-energy quark matter and low-energy hadronic matter, which renders perturbative techniques unusable, making it difficult to construct an accurate QCD-derived model of the forces between nucleons. Current approaches are limited to either phenomenological models such as the Argonne v18 potential or chiral effective field theory.

- Even if the nuclear force is well constrained, a significant amount of computational power is required to accurately compute the properties of nuclei ab initio. Developments in many-body theory have made this possible for many low mass and relatively stable nuclei, but further improvements in both computational power and mathematical approaches are required before heavy nuclei or highly unstable nuclei can be tackled.

Historically, experiments have been compared to relatively crude models that are necessarily imperfect. None of these models can completely explain experimental data on nuclear structure.

The nuclear radius (R) is considered to be one of the basic quantities that any model must predict. For stable nuclei (not halo nuclei or other unstable distorted nuclei) the nuclear radius is roughly proportional to the cube root of the mass number (A) of the nucleus, and particularly in nuclei containing many nucleons, as they arrange in more spherical configurations:

The stable nucleus has approximately a constant density and therefore the nuclear radius R can be approximated by the following formula,

$$R = r_0 A^{1/3}$$

where A = Atomic mass number (the number of protons Z, plus the number of neutrons N) and $r_0 = 1.25$ fm $= 1.25 \times 10^{-15}$ m. In this equation, the "constant" r_0 varies by 0.2 fm, depending on the nucleus in question, but this is less than 20% change from a constant.

In other words, packing protons and neutrons in the nucleus gives approximately the same total size result as packing hard spheres of a constant size (like marbles) into a tight spherical or almost spherical bag (some stable nuclei are not quite spherical, but are known to be prolate).

Models of nuclear structure include :

Liquid Drop Model

Early models of the nucleus viewed the nucleus as a rotating liquid drop. In this model, the trade-off of long-range electromagnetic forces and relatively short-range nuclear forces, together cause behavior which resembled surface tension forces in liquid drops of different sizes. This formula is successful at explaining many important phenomena of nuclei, such as their changing amounts of binding energy as their size and composition changes, but it does not explain the special stability which occurs when nuclei have special "magic numbers" of protons or neutrons.

The terms in the semi-empirical mass formula, which can be used to approximate the binding energy of many nuclei, are considered as the sum of five types of energies. Then the picture of a nucleus as a drop of incompressible liquid roughly accounts for the observed variation of binding energy of the nucleus:

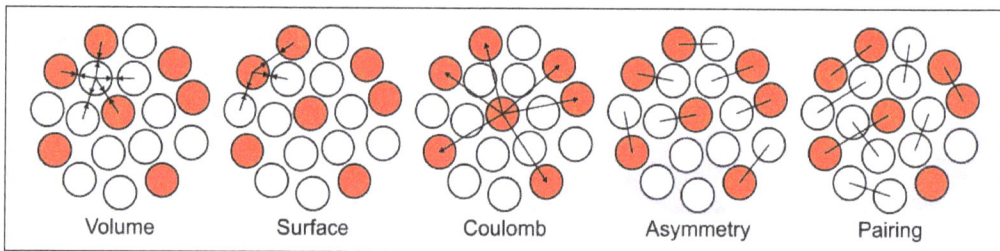

- Volume energy: When an assembly of nucleons of the same size is packed together into the smallest volume, each interior nucleon has a certain number of other nucleons in contact with it. So, this nuclear energy is proportional to the volume.

- Surface energy: A nucleon at the surface of a nucleus interacts with fewer other nucleons than one in the interior of the nucleus and hence its binding energy is less. This surface energy term takes that into account and is therefore negative and is proportional to the surface area.

- Coulomb Energy: The electric repulsion between each pair of protons in a nucleus contributes toward decreasing its binding energy.

- Asymmetry energy (also called Pauli Energy): An energy associated with the Pauli exclusion principle. Were it not for the Coulomb energy, the most stable form of nuclear matter would have the same number of neutrons as protons, since unequal numbers of neutrons and protons imply filling higher energy levels for one type of particle, while leaving lower energy levels vacant for the other type.

- Pairing energy: An energy which is a correction term that arises from the tendency of proton pairs and neutron pairs to occur. An even number of particles is more stable than an odd number.

Shell Models and other Quantum Models

A number of models for the nucleus have also been proposed in which nucleons occupy orbitals, much like the atomic orbitals in atomic physics theory. These wave models imagine nucleons to be either sizeless point particles in potential wells, or else probability waves as in the "optical model", frictionlessly orbiting at high speed in potential wells.

In the above models, the nucleons may occupy orbitals in pairs, due to being fermions, which allows explanation of even/odd Z and N effects well-known from experiments. The exact nature and capacity of nuclear shells differs from those of electrons in atomic orbitals, primarily because the potential well in which the nucleons move (especially in larger nuclei) is quite different from the central electromagnetic potential well which binds electrons in atoms. Some resemblance to atomic orbital models may be seen in a small atomic nucleus like that of helium-4, in which the two protons and two neutrons separately occupy 1s orbitals analogous to the 1s orbital for the two electrons in the helium atom, and achieve unusual stability for the same reason. Nuclei with 5 nucleons are all extremely unstable and short-lived, yet, helium-3, with 3 nucleons, is very stable even with lack of a closed 1s orbital shell. Another nucleus with 3 nucleons, the triton hydrogen-3 is unstable and will decay into helium-3 when isolated. Weak nuclear stability with 2 nucleons {NP} in the 1s orbital is found in the deuteron hydrogen-2, with only one nucleon in each of the proton and neutron potential wells. While each nucleon is a fermion, the {NP} deuteron is a boson and thus does not follow Pauli Exclusion for close packing within shells. Lithium-6 with 6 nucleons is highly stable without a closed second 1p shell orbital. For light nuclei with total nucleon numbers 1 to 6 only those with 5 do not show some evidence of stability. Observations of beta-stability of light nuclei outside closed shells indicate that nuclear stability is much more complex than simple closure of shell orbitals with magic numbers of protons and neutrons.

For larger nuclei, the shells occupied by nucleons begin to differ significantly from electron shells, but nevertheless, present nuclear theory does predict the magic numbers of filled nuclear shells for both protons and neutrons. The closure of the stable shells predicts unusually stable configurations, analogous to the noble group of nearly-inert gases in chemistry. An example is the stability of the closed shell of 50 protons, which allows tin to have 10 stable isotopes, more than any other element. Similarly, the distance from shell-closure explains the unusual instability of isotopes which have far from stable numbers of these particles, such as the radioactive elements 43 (technetium) and 61 (promethium), each of which is preceded and followed by 17 or more stable elements.

There are however problems with the shell model when an attempt is made to account for nuclear properties well away from closed shells. This has led to complex post hoc distortions of the shape of the potential well to fit experimental data, but the question remains whether these mathematical manipulations actually correspond to the spatial deformations in real nuclei. Problems with the shell model have led some to propose realistic two-body and three-body nuclear force effects involving nucleon clusters and then build the nucleus on this basis. Three such cluster models are the 1936 Resonating Group Structure model of John Wheeler, Close-Packed Spheron Model of Linus Pauling and the 2D Ising Model of MacGregor.

Consistency between Models

As with the case of superfluid liquid helium, atomic nuclei are an example of a state in which both

(1) "ordinary" particle physical rules for volume and (2) non-intuitive quantum mechanical rules for a wave-like nature apply. In superfluid helium, the helium atoms have volume, and essentially "touch" each other, yet at the same time exhibit strange bulk properties, consistent with a Bose–Einstein condensation. The nucleons in atomic nuclei also exhibit a wave-like nature and lack standard fluid properties, such as friction. For nuclei made of hadrons which are fermions, Bose-Einstein condensation does not occur, yet nevertheless, many nuclear properties can only be explained similarly by a combination of properties of particles with volume, in addition to the frictionless motion characteristic of the wave-like behavior of objects trapped in Erwin Schrödinger's quantum orbitals.

BROWNIAN MOTION

Brownian motion, also called Brownian movement is any of various physical phenomena in which some quantity is constantly undergoing small, random fluctuations. It was named for the Scottish botanist Robert Brown, the first to study such fluctuations.

If a number of particles subject to Brownian motion are present in a given medium and there is no preferred direction for the random oscillations, then over a period of time the particles will tend to be spread evenly throughout the medium. Thus, if A and B are two adjacent regions and, at time t, A contains twice as many particles as B, at that instant the probability of a particle's leaving A to enter B is twice as great as the probability that a particle will leave B to enter A. The physical process in which a substance tends to spread steadily from regions of high concentration to regions of lower concentration is called diffusion. Diffusion can therefore be considered a macroscopic manifestation of Brownian motion on the microscopic level. Thus, it is possible to study diffusion by simulating the motion of a Brownian particle and computing its average behaviour. A few examples of the countless diffusion processes that are studied in terms of Brownian motion include the diffusion of pollutants through the atmosphere, the diffusion of "holes" (minute regions in which the electrical charge potential is positive) through a semiconductor, and the diffusion of calcium through bone tissue in living organisms.

Early Investigations

The term "classical Brownian motion" describes the random movement of microscopic particles suspended in a liquid or gas. Brown was investigating the fertilization process in Clarkia pulchella, then a newly discovered species of flowering plant, when he noticed a "rapid oscillatory motion" of the microscopic particles within the pollen grains suspended in water under the microscope. Other researchers had noticed this phenomenon earlier, but Brown was the first to study it. Initially he believed that such motion was a vital activity peculiar to the male sex cells of plants, but he then checked to see if the pollen of plants dead for over a century showed the same movement. Brown called this a "very unexpected fact of seeming vitality being retained by these 'molecules' so long after the death of the plant." Further study revealed that the same motion could be observed not only with particles of other organic substances but even with chips of glass or granite and particles of smoke. Finally, in inarguable support of the nonliving nature of the phenomenon, he demonstrated it in fluid-filled vesicles in rock from the Great Sphinx.

Early explanations attributed the motion to thermal convection currents in the fluid. When observation showed that nearby particles exhibited totally uncorrelated activity, however, this simple explanation was abandoned. By the 1860s theoretical physicists had become interested in Brownian motion and were searching for a consistent explanation of its various characteristics: a given particle appeared equally likely to move in any direction; further motion seemed totally unrelated to past motion; and the motion never stopped. An experiment (1865) in which a suspension was sealed in glass for a year showed that the Brownian motion persisted. More systematic investigation in 1889 determined that small particle size and low viscosity of the surrounding fluid resulted in faster motion.

Einstein's Theory of Brownian Motion

Since higher temperatures also led to more-rapid Brownian motion, in 1877 it was suggested that its cause lay in the "thermal molecular motion in the liquid environment." The idea that molecules of a liquid or gas are constantly in motion, colliding with each other and bouncing back and forth, is a prominent part of the kinetic theory of gases developed in the third quarter of the 19th century by the physicists James Clerk Maxwell, Ludwig Boltzmann, and Rudolf Clausius in explanation of heat phenomena. According to the theory, the temperature of a substance is proportional to the average kinetic energy with which the molecules of the substance are moving or vibrating. It was natural to guess that somehow this motion might be imparted to larger particles that could be observed under the microscope; if true, this would be the first directly observable effect that would corroborate the kinetic theory. This line of reasoning led the German physicist Albert Einstein in 1905 to produce his quantitative theory of Brownian motion. Similar studies were carried out on Brownian motion, independently and almost at the same time, by the Polish physicist Marian Smoluchowski, who used methods somewhat different from Einstein's.

Einstein wrote later that his major aim was to find facts that would guarantee as much as possible the existence of atoms of definite size. In the midst of this work, he discovered that according to atomic theory there would have to be an observable movement of suspended microscopic particles. Einstein did not realize that observations concerning the Brownian motion were already long familiar. Reasoning on the basis of statistical mechanics, he showed that for such a microscopic particle the random difference between the pressure of molecular bombardment on two opposite sides would cause it to constantly wobble back and forth. A smaller particle, a less viscous fluid, and a higher temperature would each increase the amount of motion one could expect to observe. Over a period of time, the particle would tend to drift from its starting point, and, on the basis of kinetic theory, it is possible to compute the probability (P) of a particle's moving a certain distance (x) in any given direction (the total distance it moves will be greater than x) during a certain time interval (t) in a medium whose coefficient of diffusion (D) is known, D being equal to one-half the average of the square of the displacement in the x-direction. This formula for probability "density" allows P to be plotted against x. The graph is the familiar bell-shaped Gaussian "normal" curve that typically arises when the random variable is the sum of many independent, statistically identical random variables, in this case the many little pushes that add up to the total motion. The equation for this relationship is:

$$P = \frac{e^{-x^2/4Dt}}{2\sqrt{\pi Dt}}$$

The introduction of the ultramicroscope in 1903 aided quantitative studies by making visible small colloidal particles whose greater activity could be measured more easily. Several important measurements of this kind were made from 1905 to 1911. During this period the French physicist Jean-Baptiste Perrin was successful in verifying Einstein's analysis, and for this work he was awarded the Nobel Prize for Physics in 1926. His work established the physical theory of Brownian motion and ended the skepticism about the existence of atoms and molecules as actual physical entities.

HYDROGEN ATOM

The hydrogen atom is the simplest atom in nature and, therefore, a good starting point to study atoms and atomic structure. The hydrogen atom consists of a single negatively charged electron that moves about a positively charged proton. In Bohr's model, the electron is pulled around the proton in a perfectly circular orbit by an attractive Coulomb force. The proton is approximately 1800 times more massive than the electron, so the proton moves very little in response to the force on the proton by the electron. (This is analogous to the Earth-Sun system, where the Sun moves very little in response to the force exerted on it by Earth.)

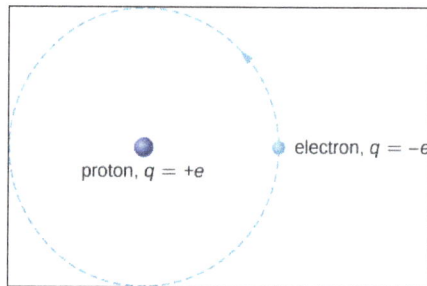

A representation of the Bohr model of the hydrogen atom.

With the assumption of a fixed proton, we focus on the motion of the electron.

In the electric field of the proton, the potential energy of the electron is:

$$U(r) = -k\frac{e^2}{r},$$

where $K = 1/4\pi\varepsilon_0$ and r is the distance between the electron and the proton. As we saw earlier, the force on an object is equal to the negative of the gradient (or slope) of the potential energy function. For the special case of a hydrogen atom, the force between the electron and proton is an attractive Coulomb force.

Notice that the potential energy function $U(r)$ does not vary in time. As a result, Schrödinger's equation of the hydrogen atom reduces to two simpler equations: one that depends only on space (x, y, z) and another that depends only on time (t). (The separation of a wave function into space- and time-dependent parts for time-independent potential energy functions is discussed in Quantum Mechanics.) We are most interested in the space-dependent equation:

$$\frac{\hbar}{2m_e}\left(\frac{\partial^2\psi}{\partial x^2} + \frac{\partial^2\psi}{\partial y^2} + \frac{\partial^2\psi}{\partial z^2}\right) - k\frac{e^2}{r}\psi = E\psi$$

where $\psi = psi\,(x,y,z)$ is the three-dimensional wave function of the electron, meme is the mass of the electron, and E is the total energy of the electron. Recall that the total wave function $\Psi = (x,y,z,t)$, is the product of the space-dependent wave function $\psi = \psi(x,y,z)$ and the time-dependent wave function $\varphi = \varphi(t)$.

In addition to being time-independent, $U(r)$ is also spherically symmetrical. This suggests that we may solve Schrödinger's equation more easily if we express it in terms of the spherical coordinates (r,θ,ϕ) instead of rectangular coordinates (x,y,z). A spherical coordinate system is shown in figure. In spherical coordinates, the variable r is the radial coordinate, θ is the polar angle (relative to the vertical z-axis), and ϕ is the azimuthal angle (relative to the x-axis). The relationship between spherical and rectangular coordinates is $x = r\sin\theta\,\cos\phi$, $y = r\sin\theta\sin\phi$, $z = r\cos\theta$.

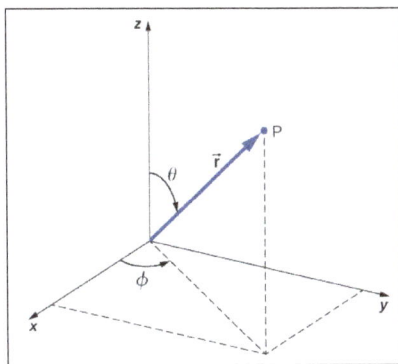

The relationship between the spherical and rectangular coordinate systems.

The factor $r\sin\theta$ is the magnitude of a vector formed by the projection of the polar vector onto the xy-plane. Also, the coordinates of x and y are obtained by projecting this vector onto the x- and y-axes, respectively. The inverse transformation gives:

$$r\sqrt{x^2+y^2+z^2},\ \theta\cos^{-1}\left(\frac{z}{r}\right),\ \phi=\cos^{-1}\left(\frac{x}{\sqrt{x^2+y^2}}\right)$$

Schrödinger's wave equation for the hydrogen atom in spherical coordinates is discussed in more advanced courses in modern physics, so we do not consider it in detail here. However, due to the spherical symmetry of $U(r)$, this equation reduces to three simpler equations: one for each of the three coordinates $(r,\theta,\text{and}\,\phi)$. Solutions to the time-independent wave function are written as a product of three functions:

$$\psi(r,\theta,\phi) = R(r)\Theta(\theta)\Phi(\phi)$$

where R is the radial function dependent on the radial coordinate r only; Θ is the polar function dependent on the polar coordinate θ only; and ϕ is the phi function of ϕ only. Valid solutions to Schrödinger's equation $\psi(r,\theta,\phi)$ are labeled by the quantum numbers n, l, and m.

n : Principal quantum number.

l: Angular momentum quantum number.

m: Angular momentum projection quantum number.

The radial function R depends only on n and l; the polar function Θ depends only on l and m ; and the phi function Φ depends only on m. The dependence of each function on quantum numbers is indicated with subscripts:

$$\psi_{nlm}(r,\theta,\phi) = R_{nl}(r)\Theta_{lm}(\theta)\Phi_m(\phi).$$

Not all sets of quantum numbers (n, l, m) are possible. For example, the orbital angular quantum number l can never be greater or equal to the principal quantum number $n(l < n)$. Specifically, we have:

$n = 1, 2, 3,...$

$l = 0, 1, 2...,(n-1)$

$m = -1,(-l+1),...,0,...,(+1-1),+1$

Notice that for the ground state, $n = 1$, $l = 0$, and $m = 0$. . In other words, there is only one quantum state with the wave function for $n = 1$, and it is ψ_{100}. However, for $n = 2$, we have:

$l = 1, m = 0$

$l = 1, m = -1,0,1.$

Therefore, the allowed states for the $n = 2$ state are ψ_{200}, ψ_{21-1}, ψ_{210} and ψ_{211} Example wave functions for the hydrogen atom are given in table. Note that some of these expressions contain the letter i, which represents $\sqrt{-1}$. When probabilities are calculated, these complex numbers do not appear in the final answer.

Table: Wave functions of the hydrogen atom.

$n = 1, l = 0, m_l = 0$	$\psi_{100} = \dfrac{1}{\sqrt{\pi}}\dfrac{1}{a_0^{3/2}}e^{-r/a_0}$
$n = 2, l = 0, m_l = 0$	$\psi_{200} = \dfrac{1}{4\sqrt{2\pi}}\dfrac{1}{a_0^{3/2}}\left(2-\dfrac{r}{a_0}\right)e^{-r/2a_0}$
$n = 2, l = 1, m_l = -1$	$\psi_{21-1} = \dfrac{1}{8\sqrt{\pi}}\dfrac{1}{a_0^{3/2}}\dfrac{r}{a_0}e^{-r2a_0}\sin\theta e^{-i\phi}$
$n = 2, l = 1, m_l = 0$	$\psi_{210} = \dfrac{1}{4\sqrt{2\pi}}\dfrac{1}{a_0^{3/2}}\dfrac{r}{a_0}e^{-r/2a_0}\cos\theta$
$n = 2, l = 1, m_l = 1$	$\psi_{211} = \dfrac{1}{8\sqrt{\pi}}\dfrac{1}{a_0^{3/2}}\dfrac{r}{a_0}e^{-r/2a_0}\sin\theta e^{i\phi}$

Physical Significance of the Quantum Numbers

Each of the three quantum numbers of the hydrogen atom (n, l, m) is associated with a different physical quantity. The principal quantum number n is associated with the total energy of the electron, E_n. According to Schrödinger's equation:

$$E_n = -\left(\frac{m_e k^2 e_4}{2^2}\right)\left(\frac{1}{n^2}\right) = -E_0\left(\frac{1}{n^2}\right),$$

where $E_0 = -13.6 \ eV$. Notice that this expression is identical to that of Bohr's model. As in the Bohr model, the electron in a particular state of energy does not radiate.

Example:

How Many Possible States?

For the hydrogen atom, how many possible quantum states correspond to the principal number $n = 3$? What are the energies of these states?

Strategy:

For a hydrogen atom of a given energy, the number of allowed states depends on its orbital angular momentum. We can count these states for each value of the principal quantum number, $n = 1, 2, 3$. However, the total energy depends on the principal quantum number only, which means that we can use Equation and the number of states counted.

Solution:

If $n = 3$, the allowed values of l are 0, 1, and 2. If $l = 0$, $m = 0$ (1 state). If $l = 1$, $m = -1, 0, 1$ (3 states); and if $l = 2$, $m = -2, -1, 0, 1, 2$ (5 states). In total, there are $1+3+5 = 9$ allowed states. Because the total energy depends only on the principal quantum number, n = 3 , the energy of each of these states is:

$$E_{n3} = -E_0\left(\frac{1}{n^2}\right) = \frac{-13.6 \ eV}{9} = -1.51 \ eV$$

Significance:

An electron in a hydrogen atom can occupy many different angular momentum states with the very same energy. As the orbital angular momentum increases, the number of the allowed states with the same energy increases.

The angular momentum orbital quantum number l is associated with the orbital angular momentum of the electron in a hydrogen atom. Quantum theory tells us that when the hydrogen atom is in the state ψ_{nlm}, the magnitude of its orbital angular momentum is:

$$L = \sqrt{l(l+1)}\hbar,$$

Where,

$$l = 0, 1, 2, ..., (n-1).$$

This result is slightly different from that found with Bohr's theory, which quantizes angular momentum according to the rule $L = n$, where $n = 1, 2, 3,...$

Quantum states with different values of orbital angular momentum are distinguished using spectroscopic notation (Table). The designations s, p, d, and f result from early historical attempts to classify atomic spectral lines. (The letters stand for sharp, principal, diffuse, and fundamental, respectively.) After f, the letters continue alphabetically.

The ground state of hydrogen is designated as the $1s$ state, where "1" indicates the energy level $(n = 1)$ and "s" indicates the orbital angular momentum state $(l = 0)$. When $n = 2$, l can be either 0 or 1. The $n = 2, l = 0$ state is designated "2s." The $n = 2, l = 0$ state is designated "2p." When $n = 3$, l can be 0, 1, or 2, and the states are $3s$, $3p$, and $3d$, respectively. Notation for other quantum states is given in table.

The angular momentum projection quantum number m is associated with the azimuthal angle ϕ and is related to the z-component of orbital angular momentum of an electron in a hydrogen atom. This component is given by,

$$L_z = m\hbar,$$

where,

$$m = -l, -l+1, ...0, ..., +l-l, l.$$

The z-component of angular momentum is related to the magnitude of angular momentum by,

$$L_z = L\cos\theta,$$

where θ is the angle between the angular momentum vector and the z-axis. Note that the direction of the z-axis is determined by experiment - that is, along any direction, the experimenter decides to measure the angular momentum. For example, the z-direction might correspond to the direction of an external magnetic field. The relationship between L_z and L is given in figure.

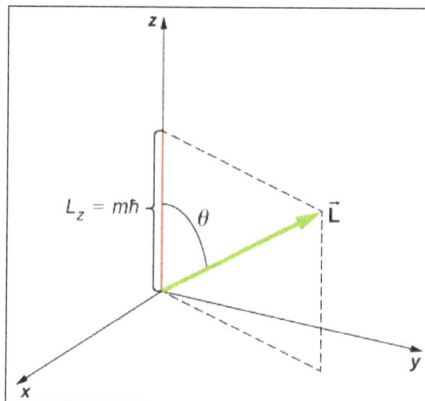

The z-component of angular momentum is quantized with its own quantum number m.

Table: Spectroscopic notation and orbital angular momentum.

Orbital Quantum Number l	Angular Momentum	State	Spectroscopic Name
0	0	s	Sharp
1	$\sqrt{2}h$	p	Principal
2	$\sqrt{6}h$	d	Diffuse
3	$\sqrt{12}h$	f	Fundamental
4	$\sqrt{20}h$	g	
5	$\sqrt{30}h$	h	

Table: Spectroscopic description of quantum states.

	$l=1$	$l=2$	$l=3$	$l=3$	$l=4$	$l=5$
$n=1$	$1s$					
$n=2$	$2s$	$2p$				
$n=3$	$3s$	$3p$	$3d$			
$n=4$	$4s$	$4p$	$4d$	$4f$		
$n=5$	$5s$	$5p$	$5d$	$5f$	$5g$	
$n=6$	$6s$	$6p$	$6d$	$6f$	$6g$	$6h$

The quantization of L_z is equivalent to the quantization of $|$ *theta*. Substituting $\sqrt{l(l+1)}\hbar$ L and for L_z into this equation, we find:

$$m\hbar = \sqrt{l(l+1)}\hbar \, \cos\theta.$$

Thus, the angle θ is quantized with the particular values:

$$\theta = \cos^{-1}\left(\frac{m}{\sqrt{l(l+1)}}\right).$$

Notice that both the polar angle $(\theta\theta)$ and the projection of the angular momentum vector onto an arbitrary z-axis (L_z) are quantized.

The quantization of the polar angle for the $l=3$ state is shown in figure. The orbital angular momentum vector lies somewhere on the surface of a cone with an opening angle θ relative to the z-axis (unless $m=0$, in which case $\theta=90°$ and the vector points are perpendicular to the z-axis).

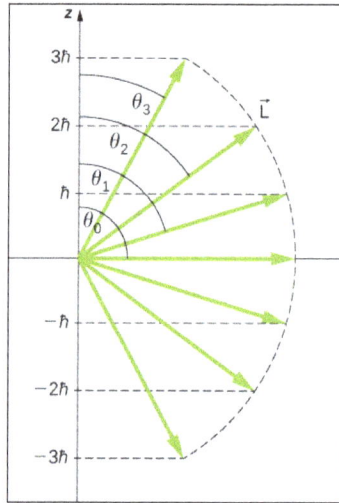

The quantization of orbital angular momentum. Each vector lies
on the surface of a cone with axis along the z-axis.

A detailed study of angular momentum reveals that we cannot know all three components simultaneously. The z-component of orbital angular momentum has definite values that depend on the quantum number m. This implies that we cannot know both x- and y-components of angular momentum, L_x and L_y, with certainty. As a result, the precise direction of the orbital angular momentum vector is unknown.

Example:

What are the allowed directions?

Calculate the angles that the angular momentum vector $\vec{L}$ can make with the z-axis for $l = 1$, as shown in figure.

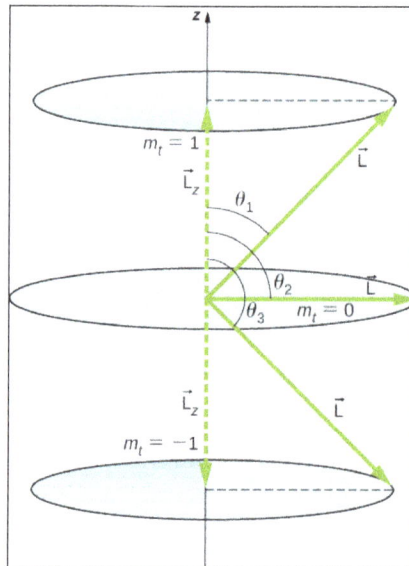

The component of a given angular momentum along the z-axis (defined by the direction of a magnetic field) can have only certain values. These are shown here for $l = 1$, for which $m = -1$, 0 and + 1.
The direction of $\vec{L}$ is quantized in the sense that it can have only certain angles relative to the z-axis.

Strategy:

The vectors $\vec{L}$ and $\vec{L_z}$ (in the z-direction) form a right triangle, where $\vec{L}$ is the hypotenuse and $\vec{L_z}$ is the adjacent side. The ratio of $\vec{L_z}$ to $|\vec{L}|$ is the cosine of the angle of interest. The magnitudes $L = |\vec{L}|$ and $\vec{L_z}$ are given by:

$$L = \sqrt{l(l+1)}\hbar \text{ and } L_z = m\hbar.$$

Solution:

We are given $l = 1$, so ml can be +1, 0, or +1. Thus, L has the value given by:

$$L = \sqrt{l(l+1)}\hbar = \sqrt{2}\hbar$$

The quantity L_z can have three values, given by $L_z = m_l\hbar$.

$$L_z = m_l\hbar = \hbar, \ m_l = +1$$

$$= 0, \ m_l = 0$$

$$= -\hbar, \ m_l = -1$$

As you can see in figure, $\cos\theta = lz / L, \cos\theta = Lz / L$, so for $m = +1m = +1$, we have:

$$\cos\theta_1 = \frac{L_z}{L} = \frac{\hbar}{\sqrt{2}\hbar} = \frac{1}{\sqrt{2}} = 0.707$$

Thus,

$$\theta_1 = \cos^{-1} 0.707 = 45.0°.$$

Then for $ml = -1$:

$$\cos\theta_3 = \frac{L_z}{L} = \frac{-\hbar}{\sqrt{2}\hbar} = -\frac{1}{\sqrt{2}} = -0.707$$

so that:

$$\theta_3 \cos^{-1}(-0.707) = 135.0°.$$

Significance:

The angles are consistent with the figure. Only the angle relative to the z-axis is quantized. L can point in any direction as long as it makes the proper angle with the z-axis. Thus, the angular momentum vectors lie on cones, as illustrated. To see how the correspondence principle holds here, consider that the smallest angle (θ_1 in the example) is for the maximum value of ml, ml, namely $m_l = l$. For that smallest angle,

$$\cos\theta = \frac{L_z}{L} = \frac{1}{\sqrt{l(l+1)}},$$

which approaches 1 as l becomes very large. If $\cos\theta = 1$, then $\theta = 0°$. Furthermore, for large l, there are many values of m_l, so that all angles become possible as l gets very large.

Using the Wave Function to Make Predictions

As we saw earlier, we can use quantum mechanics to make predictions about physical events by the use of probability statements. It is therefore proper to state, "An electron is located within this volume with this probability at this time," but not, "An electron is located at the position (x, y, z) at this time." To determine the probability of finding an electron in a hydrogen atom in a particular region of space, it is necessary to integrate the probability density $|\psi nlm|2|\psi nlm|2$ over that region:

$$Probability = \int_{volume} \left|\psi_{nlm}\right|^2 dV,$$

where dV is an infinitesimal volume element. If this integral is computed for all space, the result is 1, because the probability of the particle to be located somewhere is 100% (the normalization condition). In a more advanced course on modern physics, you will find that $\left|\psi_{nlm}\right|^2 = \psi_{nlm}^* \psi_{nlm}$, where ψ_{nlm}^* is the complex conjugate. This eliminates the occurrences $i = \sqrt{-1}$ in the above calculation.

Consider an electron in a state of zero angular momentum $(l = 0)$. In this case, the electron's wave function depends only on the radial coordinate r. The infinitesimal volume element corresponds to a spherical shell of radius r and infinitesimal thickness dr, written as:

$$dV = 4\pi r^2 dr.$$

The probability of finding the electron in the region r to $r + dr$ ("at approximately r") is:

$$P(r)dr = \left|\psi_{n00}\right|^2 4\pi r^2 dr.$$

Here $P(r)$ is called the radial probability density function (a probability per unit length). For an electron in the ground state of hydrogen, the probability of finding an electron in the region r to $r + dr$ is:

$$\left|\psi_{n00}\right|^2 4\pi r^2 dr = 4\left(4 / a_0^3\right) r^2 \exp\left(-2r / a_0\right) dr,$$

where $a_0 = 0.5$ angstroms. The radial probability density function $P(r)$ is plotted in figure. The area under the curve between any two radial positions, say r_1 and r_2, gives the probability of finding the electron in that radial range. To find the most probable radial position, we set the first derivative of this function to zero $(dP / dr = 0)$ and solve for r. The most probable radial position is not equal to the average or expectation value of the radial position because $\left|\psi_{n00}\right|^2$ s not symmetrical about its peak value.

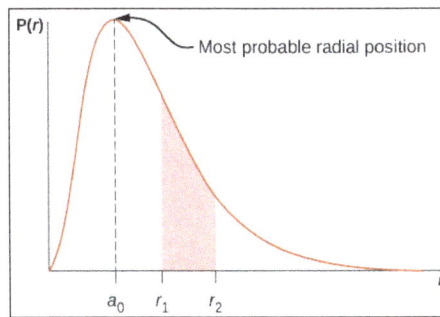

The radial probability density function for the ground state of hydrogen.

If the electron has orbital angular momentum $(l \neq 0)$, then the wave functions representing the electron depend on the angles θ and ϕ; that is, $\psi_{nlm} = \psi_{nlm}(r, \theta, \phi)$. Atomic orbitals for three states with $n = 2$ and $l = 1$ are shown in figure. An atomic orbital is a region in space that encloses a certain percentage (usually 90%) of the electron probability. (Sometimes atomic orbitals are referred to as "clouds" of probability.) Notice that these distributions are pronounced in certain directions. This directionality is important to chemists when they analyze how atoms are bound together to form molecules.

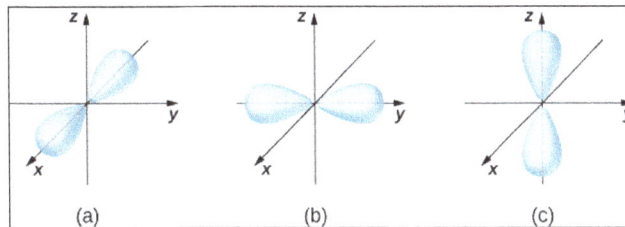

The probability density distributions for three states with n = 2n = 2 and l = 1l = 1. The distributions are directed along the (a) x-axis, (b) y-axis, and (c) z-axis.

A slightly different representation of the wave function is given in Figure. In this case, light and dark regions indicate locations of relatively high and low probability, respectively. In contrast to the Bohr model of the hydrogen atom, the electron does not move around the proton nucleus in a well-defined path. Indeed, the uncertainty principle makes it impossible to know how the electron gets from one place to another.

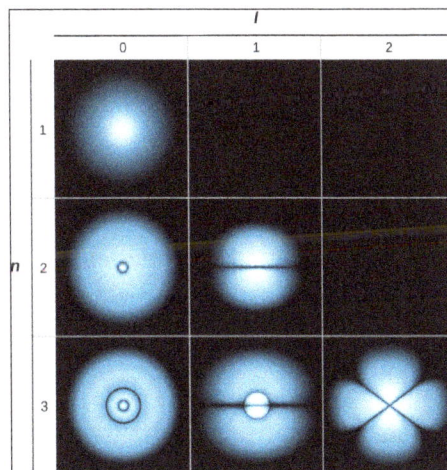

Probability clouds for the electron in the ground state and several excited states of hydrogen. The probability of finding the electron is indicated by the shade of color; the lighter the coloring, the greater the chance of finding the electron.

Isotopes

The most abundant isotope, hydrogen-1, protium, or light hydrogen, contains no neutrons and is simply a proton and an electron. Protium is stable and makes up 99.985% of naturally occurring hydrogen atoms.

Deuterium contains one neutron and one proton. Deuterium is stable and makes up 0.0156% of naturally occurring hydrogen and is used in industrial processes like nuclear reactors and Nuclear Magnetic Resonance.

Tritium contains two neutrons and one proton and is not stable, decaying with a half-life of 12.32 years. Because of the short half life, tritium does not exist in nature except in trace amounts.

Higher isotopes of hydrogen are only created in artificial accelerators and reactors and have half lives around the order of 10^{-22} second. They are unbound resonances located beyond the neutron drip line; this results in prompt emission of a neutron.

The formulas below are valid for all three isotopes of hydrogen, but slightly different values of the Rydberg constant (correction formula given below) must be used for each hydrogen isotope.

Hydrogen Ion

Lone neutral hydrogen atoms are rare under normal conditions. However, neutral hydrogen is common when it is covalently bound to another atom, and hydrogen atoms can also exist in cationic and anionic forms.

If a neutral hydrogen atom loses its electron, it becomes a cation. The resulting ion, which consists solely of a proton for the usual isotope, is written as "H^+" and sometimes called *hydron*. Free protons are common in the interstellar medium, and solar wind. In the context of aqueous solutions of classical Brønsted–Lowry acids, such as hydrochloric acid, it is actually hydronium, H_3O^+, that is meant. Instead of a literal ionized single hydrogen atom being formed, the acid transfers the hydrogen to H_2O, forming H_3O^+.

If instead a hydrogen atom gains a second electron, it becomes an anion. The hydrogen anion is written as "H^-" and called hydride.

Theoretical Analysis

The hydrogen atom has special significance in quantum mechanics and quantum field theory as a simple two-body problem physical system which has yielded many simple analytical solutions in closed-form.

Failed Classical Description

Experiments by Ernest Rutherford in 1909 showed the structure of the atom to be a dense, positive nucleus with a tenuous negative charge cloud around it. This immediately raised questions about how such a system could be stable. Classical electromagnetism had shown that any accelerating charge radiates energy, as shown by the Larmor formula. If the electron is assumed to orbit in a perfect circle and radiates energy continuously, the electron would rapidly spiral into the nucleus with a fall time of:

$$t_{fall} \approx \frac{a_0^3}{4r_0^2 c} \approx~ 1.6 . 10^{-11} s$$

Where a_0 is the Bohr radius and r_0 is the classical electron radius. If this were true, all atoms would instantly collapse, however atoms seem to be stable. Furthermore, the spiral inward would release a smear of electromagnetic frequencies as the orbit got smaller. Instead, atoms were observed to only emit discrete frequencies of radiation. The resolution would lie in the development of quantum mechanics.

Bohr–Sommerfeld Model

In 1913, Niels Bohr obtained the energy levels and spectral frequencies of the hydrogen atom after making a number of simple assumptions in order to correct the failed classical model. The assumptions included:

- Electrons can only be in certain, discrete circular orbits or stationary states, thereby having a discrete set of possible radii and energies.

- Electrons do not emit radiation while in one of these stationary states.

- An electron can gain or lose energy by jumping from one discrete orbital to another.

Bohr supposed that the electron's angular momentum is quantized with possible values:

$$L = n\hbar \text{ where } n = 1, 2, 3,...$$

And $\hbar$ is Planck constant over 2π. He also supposed that the centripetal force which keeps the electron in its orbit is provided by the Coulomb force, and that energy is conserved. Bohr derived the energy of each orbit of the hydrogen atom to be:

$$E_n = -\frac{m_e e^4}{2(4\pi\varepsilon_0)^2 \hbar^2} \frac{1}{n^2}$$

Where m_e is the electron mass, e is the electron charge, ε_0 is the vacuum permittivity, and n is the quantum number (now known as the principal quantum number). Bohr's predictions matched experiments measuring the hydrogen spectral series to the first order, giving more confidence to a theory that used quantized values.

For $n = 1$, the value

$$\frac{m_e e^4}{2(4\pi\varepsilon_0)^2 \hbar^2} = \frac{m_e e^4}{8\hbar^2 \varepsilon_0^2} = 1\,\text{Ry} = 13.605\ 692\ 53(30)\ \text{eV}$$

is called the Rydberg unit of energy. It is related to the Rydberg constant R_∞ of atomic physics by $1\,\text{Ry} = hcR_\infty$.

The exact value of the Rydberg constant assumes that the nucleus is infinitely massive with respect to the electron. For hydrogen-1, hydrogen-2 (deuterium), and hydrogen-3 (tritium) which have finite mass, the constant must be slightly modified to use the reduced mass of the system, rather than simply the mass of the electron. This includes the kinetic energy of the nucleus in the problem, because the total (electron plus nuclear) kinetic energy is equivalent to the kinetic energy of the reduced mass moving with a velocity equal to the electron velocity relative to the nucleus. However, since the nucleus is much heavier than the electron, the electron mass and reduced mass are nearly the same. The Rydberg constant R_M for a hydrogen atom (one electron), R is given by:

$$R_M = \frac{R_\infty}{1 + m_e / M},$$

Where M is the mass of the atomic nucleus. For hydrogen-1, the quantity m_e / M, is about 1/1836 (i.e. the electron-to-proton mass ratio). For deuterium and tritium, the ratios are about 1/3670 and 1/5497 respectively. These figures, when added to 1 in the denominator, represent very small corrections in the value of R, and thus only small corrections to all energy levels in corresponding hydrogen isotopes.

There were still problems with Bohr's model:

- It failed to predict other spectral details such as fine structure and hyperfine structure.

- It could only predict energy levels with any accuracy for single–electron atoms (hydrogen–like atoms).

- The predicted values were only correct to $\propto^2 \approx \sim 10^{-5}$, where $\propto$ is the fine-structure constant.

Most of these shortcomings were resolved by Arnold Sommerfeld's modification of the Bohr model. Sommerfeld introduced two additional degrees of freedom, allowing an electron to move on an elliptical orbit characterized by its eccentricity and declination with respect to a chosen axis. This introduced two additional quantum numbers, which correspond to the orbital angular momentum and its projection on the chosen axis. Thus the correct multiplicity of states (except for the factor 2 accounting for the yet unknown electron spin) was found. Further, by applying special relativity to the elliptic orbits, Sommerfeld succeeded in deriving the correct expression for the fine structure of hydrogen spectra (which happens to be exactly the same as in the most elaborate Dirac theory). However, some observed phenomena, such as the anomalous Zeeman effect, remained unexplained. These issues were resolved with the full development of quantum mechanics and the Dirac equation. It is often alleged that the Schrödinger equation is superior to the Bohr–Sommerfeld theory in describing hydrogen atom. This is not the case, as most of the results of both approaches coincide or are very close (a remarkable exception is the problem of hydrogen atom in crossed electric and magnetic fields, which cannot be self-consistently solved in the framework of the Bohr–Sommerfeld theory), and in both theories the main shortcomings result from the absence of the electron spin. It was the complete failure of the Bohr–Sommerfeld theory to explain many-electron systems (such as helium atom or hydrogen molecule) which demonstrated its inadequacy in describing quantum phenomena.

Schrödinger Equation

The Schrödinger equation allows one to calculate the stationary states and also the time evolution of quantum systems. Exact analytical answers are available for the nonrelativistic hydrogen atom. Before we go to present a formal account, here we give an elementary overview.

Given that the hydrogen atom contains a nucleus and an electron, quantum mechanics allows one to predict the probability of finding the electron at any given radial distance r It is given by the square of a mathematical function known as the "wavefunction," which is a solution of the Schrödinger equation. The lowest energy equilibrium state of the hydrogen atom is known as the ground state. The ground state wave function is known as the $1s$ wavefunction. It is written as:

$$\psi 1_s (r) = \frac{1}{\sqrt{\pi a_0^{3/2}}} e^{-2r/a_0}.$$

Here, a_0 is the numerical value of the Bohr radius. The probability of finding the electron at a distance r in any radial direction is the squared value of the wavefunction:

$$\left| \psi 1_s (r) \right|^2 = \frac{1}{\pi a_0^3} e^{-2ra_0}$$

The $1s$ wavefunction is spherically symmetric, and the surface area of a shell at distance r is $4\pi r^2$, so the total probability $P(r)dr$ of the electron being in a shell at a distance r and thickness dr is:

$$P(r)dr = 4\pi r^2 \left| \psi 1_s (r) \right|^2 dr$$

It turns out that this is a maximum at $r = a_0$ That is, the Bohr picture of an electron orbiting the nucleus at radius a_0 is recovered as a statistically valid result. However, although the electron is most likely to be on a Bohr orbit, there is a finite probability that the electron may be at any other place r, with the probability indicated by the square of the wavefunction. Since the probability of finding the electron somewhere in the whole volume is unity, the integral of $P(r)dr$ is unity. Then we say that the wavefunction is properly normalized.

As discussed below, the ground state $1s$ is also indicated by the quantum numbers $n = 1$, $l = 0$, $m = 0$ The second lowest energy states, just above the ground state, are given by the quantum numbers $n - 2$, $l = 0$, $m - 0$; $n = 2$, $l = 1$, $m = 0$; and $n = 2$, $l = 1$, $m = \pm 1$. These $n = 2$ states all have the same energy and are known as the $2s$ and $2p$ states. There is one $2s$ state:

$$\psi_{2,0,0} = \frac{1}{4\sqrt{2\pi}a_0^{3/2}} \left(2 - \frac{r}{a_0} \right) e^{-r/2a_0},$$

and there are three $2p$ states:

$$\psi_{2,1,0} = \frac{1}{4\sqrt{2\pi}a_0^{3/2}} \frac{r}{a_0} e^{-r/2a_0} \cos\theta$$

$$\psi_{2,1,\pm 1} = \frac{1}{8\sqrt{\pi}a_0^{3/2}} \frac{r}{a_0} e^{-r/2a_0} \sin\theta\, e^{\pm i\phi}$$

An electron in the $2s$ or $2p$ state is most likely to be found in the second Bohr orbit with energy given by the Bohr formula.

Wavefunction

The Hamiltonian of the hydrogen atom is the radial kinetic energy operator and coulomb attraction force between the positive proton and negative electron. Using the time-independent Schrödinger equation, ignoring all spin-coupling interactions and using the reduced mass $\mu = m_e M / (m_e + M)$, the equation is written as:

$$\left(-\frac{\hbar^2}{2\mu} \nabla^2 - \frac{e^2}{4\pi\varepsilon_0 r} \right) \psi(r,\theta,\phi) = E\psi(r,\theta,\phi)$$

Expanding the Laplacian in spherical coordinates:

$$-\frac{\hbar^2}{2\mu} \left[\frac{1}{r^2} \frac{\partial}{\partial r} \left(r^2 \frac{\partial \psi}{\partial r} \right) + \frac{1}{r^2 \sin\theta} \frac{\partial}{\partial \theta} \left(\sin\theta \frac{\partial \psi}{\partial \theta} \right) + \frac{1}{r^2 \sin^2\theta} \frac{\partial^2 \psi}{\partial \phi^2} \right] - \frac{e^2}{4\pi\varepsilon_0 r} \psi = E\psi$$

This is a separable, partial differential equation which can be solved in terms of special functions. The normalized position wavefunctions, given in spherical coordinates are:

$$\psi_{nlm}(r,\vartheta,\varphi) = \sqrt{\left(\frac{2}{na_0^*} \right)^3 \frac{(n-\ell-1)!}{2n(n+\ell)!}} e^{-\rho/2} \rho^\ell L_{n-\ell-1}^{2\ell+1}(\rho) Y_\ell^m(\vartheta,\varphi)$$

where:

$$\rho = \frac{2r}{na_0^*},$$

a_0^* is the reduced Bohr radius, $a_0^* = \dfrac{4\pi\varepsilon\hbar^2}{\mu e^2}$

$L_{n-\ell-1}^{2\ell+1}(\rho)$ is a generalized Laguerre polynomial of degree $n-\ell-1$,

$Y_\ell^m(\vartheta,\varphi)$ is a spherical harmonic function of degree ℓ and order m. Note that the generalized Laguerre polynomials are defined differently by different authors. The usage here is consistent with the definitions used by Messiah, and Mathematica. In other places, the Laguerre polynomial includes a factor of $(n+\ell)!$, or the generalized Laguerre polynomial appearing in the hydrogen wave function is $L_{n+\ell}^{2\ell+1}(\rho)$ instead.

The quantum numbers can take the following values:

$n = 1, 2, 3, ...$

$\ell = 0, 1.2, 3, ... n-1$

$m = -\ell, ..., \ell$

Additionally, these wavefunctions are normalized (i.e., the integral of their modulus square equals 1) and orthogonal:

$$\int_0^\infty r^2 dr \int_0^\pi \sin\vartheta\, d\vartheta \int_0^{2\pi} d\varphi\, \psi_{n\ell m}^* (r,\vartheta,\varphi) = \langle n,\ell,m | n',\ell',m' \rangle = \delta_{nn'}\, \delta_{\ell\ell'}\, \delta_{mm'}$$

where $|n,\ell,m\rangle$ is the state represented by the wavefunction $\psi_{n\ell m}$ in Dirac notation, and δ is the Kronecker delta function.

The wavefunctions in momentum space are related to the wavefunctions in position space through a Fourier transform:

$$\phi\left(p,\vartheta_p,\varphi_p\right) = \pm\left(2\pi\hbar\right)^{-3/2} \int e^{-i\vec{p}\cdot\vec{r}/\hbar} \psi\left(r,\vartheta,\varphi\right) dV$$

which, for the bound states, results in:

$$\phi\left(p,\vartheta_p,\varphi_p\right) = \sqrt{\frac{2}{\pi}\frac{(n-l-1)!}{(n+1)!}} n^2 2^{2l+2} l! \frac{n^l p^l}{\left(n^2 p^2 +1\right)^{l+2}} C_{n-l-1}^{l+1}\left(\frac{n^2 p^2 -1}{n^2 p^2 +1}\right) Y_l^m\left(\vartheta_p,\varphi_p\right),$$

where $C_N^\alpha\left(x\right)$ denotes a Gegenbauer polynomial and p is in units of $\hbar/a_0^*$

The solutions to the Schrödinger equation for hydrogen are analytical, giving a simple expression for the hydrogen energy levels and thus the frequencies of the hydrogen spectral lines and fully reproduced the Bohr model and went beyond it. It also yields two other quantum numbers and the shape of the electron's wave function ("orbital") for the various possible quantum-mechanical states, thus explaining the anisotropic character of atomic bonds.

The Schrödinger equation also applies to more complicated atoms and molecules. When there is more than one electron or nucleus the solution is not analytical and either computer calculations are necessary or simplifying assumptions must be made.

Since the Schrödinger equation is only valid for non-relativistic quantum mechanics, the solutions it yields for the hydrogen atom are not entirely correct. The Dirac equation of relativistic quantum theory improves these solutions.

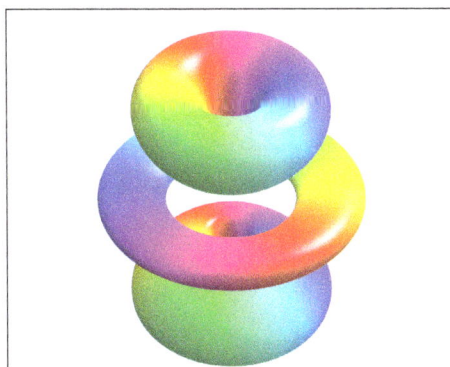

3D illustration of the eigenstate $\psi_{4,3,1}$. Electrons in this state are 45% likely to be found within the solid body shown.

Results of Schrödinger Equation

The solution of the Schrödinger equation (wave equation) for the hydrogen atom uses the fact that the Coulomb potential produced by the nucleus is isotropic (it is radially symmetric in space and only depends on the distance to the nucleus). Although the resulting energy eigenfunctions (the orbitals) are not necessarily isotropic themselves, their dependence on the angular coordinates follows completely generally from this isotropy of the underlying potential: the eigenstates of the Hamiltonian (that is, the energy eigenstates) can be chosen as simultaneous eigenstates of the angular momentum operator. This corresponds to the fact that angular momentum is conserved in the orbital motion of the electron around the nucleus. Therefore, the energy eigenstates may be classified by two angular momentum quantum numbers, ℓ and m (both are integers). The angular momentum quantum number $\ell = 0, 1, 2, \ldots$ determines the magnitude of the angular momentum. The magnetic quantum number $m = -\ell, \ldots + \ell$ determines the projection of the angular momentum on the (arbitrarily chosen) z-axis.

In addition to mathematical expressions for total angular momentum and angular momentum projection of wavefunctions, an expression for the radial dependence of the wave functions must be found. It is only here that the details of the 1/r Coulomb potential enter (leading to Laguerre polynomials in r). This leads to a third quantum number, the principal quantum number $n = 1, 2, 3, \ldots$. The principal quantum number in hydrogen is related to the atom's total energy.

Note that the maximum value of the angular momentum quantum number is limited by the principal quantum number: it can run only up to $n - 1$, i.e. $\ell = 0, 1, \ldots, n - 1$.

Due to angular momentum conservation, states of the same ℓ but different m have the same energy (this holds for all problems with rotational symmetry). In addition, for the hydrogen atom, states of the same   but different ℓ are also degenerate (i.e. they have the same energy). However, this is a specific property of hydrogen and is no longer true for more complicated atoms which have an (effective) potential differing from the form 1/r (due to the presence of the inner electrons shielding the nucleus potential).

Taking into account the spin of the electron adds a last quantum number, the projection of the electron's spin angular momentum along the z-axis, which can take on two values. Therefore, any eigenstate of the electron in the hydrogen atom is described fully by four quantum numbers. According to the usual rules of quantum mechanics, the actual state of the electron may be any superposition of these states. This explains also why the choice of z-axis for the directional quantization of the angular momentum vector is immaterial: an orbital of given ℓ and m' obtained for another preferred axis z' can always be represented as a suitable superposition of the various states of different m (but same l) that have been obtained for z.

Mathematical Summary of Eigenstates of Hydrogen Atom

In 1928, Paul Dirac found an equation that was fully compatible with Special Relativity, and (as a consequence) made the wave function a 4-component "Dirac spinor" including "up" and "down" spin components, with both positive and "negative" energy (or matter and antimatter). The solution to this equation gave the following results, more accurate than the Schrödinger solution.

Energy Levels

The energy levels of hydrogen, including fine structure (excluding Lamb shift and hyperfine structure), are given by the Sommerfeld fine structure expression:

$$E_{jn} = -\mu c^2 \left[1 - \left(1 + \left[\frac{\alpha}{n - j - \frac{1}{2} + \sqrt{\left(j + \frac{1}{2}\right)^2 - \alpha^2}} \right]^2 \right)^{-1/2} \right]$$

$$\approx \sim -\frac{\mu c^2 \alpha^2}{2n^2} \left[1 + \frac{\alpha^2}{n^2} \left(\frac{n}{j + \frac{1}{2}} - \frac{3}{4} \right) \right],$$

where α is the fine-structure constant and j is the "total angular momentum" quantum number, which is equal to $\left| \ell \pm \frac{1}{2} \right|$ depending on the direction of the electron spin. This formula represents a small correction to the energy obtained by Bohr and Schrödinger as given above. The factor in square brackets in the last expression is nearly one; the extra term arises from relativistic effects. It is worth noting that this expression was first obtained by A. Sommerfeld in 1916 based on the relativistic version of the old Bohr theory. Sommerfeld has however used different notation for the quantum numbers.

Coherent States

The coherent states have been proposed as:

$$\left| s, \gamma, \overline{\Omega} \right\rangle \equiv {}^\wedge M\left(s^2\right) \sum_{n=0}^{\infty} \left(s^n e^{i\gamma/(n+1)^2} / \sqrt{\rho_n} \right) \left| n, \overline{\Omega} \right\rangle$$

which satisfies $d\overline{\Omega} \equiv \sin \overline{\theta} d \overline{\theta} d \overline{\phi} d \overline{\psi} / 8\pi$ and takes the form:

$$\left\langle r\theta\phi \mid s, \gamma, \overline{\Omega} \right\rangle = e^{-s2/2} \sum_{n=0}^{\infty} \left(s^n e^{i\gamma/(n+1)^2} / \sqrt{n!} \right)$$

$$\times \sum_{l=0}^{n} u_{n+1}^{l}(r) \sum_{m=-l}^{l} \left[\frac{(2l)!}{(l+m)!\,(l-m)!} \right]^{1/2} \left(\sin \frac{\overline{\theta}}{2} \right)^{l-m} \left(\cos \frac{\overline{\theta}}{2} \right)^{l+m}$$

$$\times e^{-i\left(m\overline{\phi} + l\overline{\psi}\right)} Y_{lm}(\theta,\phi) \sqrt{2l+1}$$

Visualizing the Hydrogen Electron Orbitals

The image below shows the first few hydrogen atom orbitals (energy eigenfunctions). These are cross-sections of the probability density that are color-coded (black represents zero density and white represents the highest density). The angular momentum (orbital) quantum number ℓ is denoted in each column, using the usual spectroscopic letter code (s means $\ell = 0$, p means $\ell = 1$, d means $\ell = 2$. The main (principal) quantum number n (= 1, 2, 3, ...) is marked to the right of each row. For all pictures the magnetic quantum number m has been set to 0, and the cross-sectional plane is the xz-plane (z is the vertical axis). The probability density in three-dimensional space is obtained by rotating the one shown here around the z-axis.

The "ground state", i.e. the state of lowest energy, in which the electron is usually found, is the first one, the 1 s state (principal quantum level $n = 1$, $t = 0$).

Black lines occur in each but the first orbital: these are the nodes of the wavefunction, i.e. where the probability density is zero. (More precisely, the nodes are spherical harmonics that appear as a result of solving Schrödinger equation in polar coordinates.)

The quantum numbers determine the layout of these nodes. There are:

- $n - 1$ total nodes,

- l of which are angular nodes:

    ○ m angular nodes go around the ϕ axis (in the xy plane). (The figure above does not show these nodes since it plots cross-sections through the xz-plane.)

    ○ $l - m$ (the remaining angular nodes) occur on the θ (vertical) axis.

- $n - l - 1$ (the remaining non-angular nodes) are radial nodes.

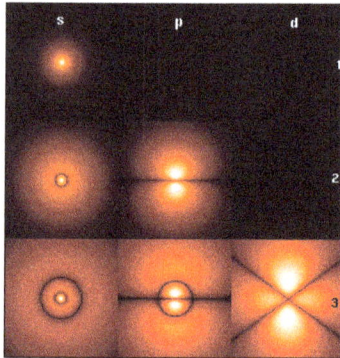

Probability densities through the xz-plane for the electron at different quantum numbers (ℓ, across top; n, down side; m = 0).

Features going Beyond the Schrödinger Solution

There are several important effects that are neglected by the Schrödinger equation and which are responsible for certain small but measurable deviations of the real spectral lines from the predicted ones.

- Although the mean speed of the electron in hydrogen is only 1/137th of the speed of light, many modern experiments are sufficiently precise that a complete theoretical explanation requires a fully relativistic treatment of the problem. A relativistic treatment results in a momentum increase of about 1 part in 37,000 for the electron. Since the electron's wavelength is determined by its momentum, orbitals containing higher speed electrons show contraction due to smaller wavelengths.

- Even when there is no external magnetic field, in the inertial frame of the moving electron, the electromagnetic field of the nucleus has a magnetic component. The spin of the electron has an associated magnetic moment which interacts with this magnetic field. This effect is also explained by special relativity, and it leads to the so-called spin-orbit coupling, i.e., an interaction between the electron's orbital motion around the nucleus, and its spin.

Both of these features (and more) are incorporated in the relativistic Dirac equation, with predictions that come still closer to experiment. Again the Dirac equation may be solved analytically in the special case of a two-body system, such as the hydrogen atom. The resulting solution quantum states now must be classified by the total angular momentum number j (arising through the coupling between electron spin and orbital angular momentum). States of the same j and the same n are still degenerate. Thus, direct analytical solution of Dirac equation predicts $2S\left(\dfrac{1}{2}\right)$ and $2P\left(\dfrac{1}{2}\right)$ levels of Hydrogen to have exactly the same energy, which is in a contradiction with observations (Lamb-Retherford experiment).

- There are always vacuum fluctuations of the electromagnetic field, according to quantum mechanics. Due to such fluctuations degeneracy between states of the same j but different l is lifted, giving them slightly different energies. This has been demonstrated in the famous Lamb-Retherford experiment and was the starting point for the development of the theory of Quantum electrodynamics (which is able to deal with these vacuum fluctuations and employs the famous Feynman diagrams for approximations using perturbation theory). This effect is now called Lamb shift.

For these developments, it was essential that the solution of the Dirac equation for the hydrogen atom could be worked out exactly, such that any experimentally observed deviation had to be taken seriously as a signal of failure of the theory.

Alternatives to the Schrödinger Theory

In the language of Heisenberg's matrix mechanics, the hydrogen atom was first solved by Wolfgang Pauli using a rotational symmetry in four dimensions [O(4)-symmetry] generated by the angular momentum and the Laplace–Runge–Lenz vector. By extending the symmetry group O(4) to the dynamical group O(4,2), the entire spectrum and all transitions were embedded in a single irreducible group representation.

In 1979 the (non relativistic) hydrogen atom was solved for the first time within Feynman's path integral formulation of quantum mechanics. This work greatly extended the range of applicability of Feynman's method.

ISOTOPES OF HYDROGEN

Hydrogen $\left(_1H\right)$ has three naturally occurring isotopes, sometimes denoted $^1H, ^2H,$ and ^3H. The first two of these are stable, while ^3H has a half-life of 12.32 years. There are also heavier isotopes, which are all synthetic and have a half-life less than one zeptosecond ($10-21$ second). Of these, ^5H is the most stable, and ^7H is the least.

Hydrogen is the only element whose individual isotopes have different names in common use today: the ^2H (or hydrogen-2) isotope is deuterium and the ^3H (or hydrogen-3) isotope is tritium. The symbols D and T are sometimes used for deuterium and tritium. The IUPAC accepts the D and T symbols, but recommends instead using standard isotopic symbols $\left(^2H \ and ^3H\right)$ to avoid confusion in the alphabetic sorting of chemical formulas. The ordinary isotope of hydrogen, with no neutrons, is sometimes called "protium".

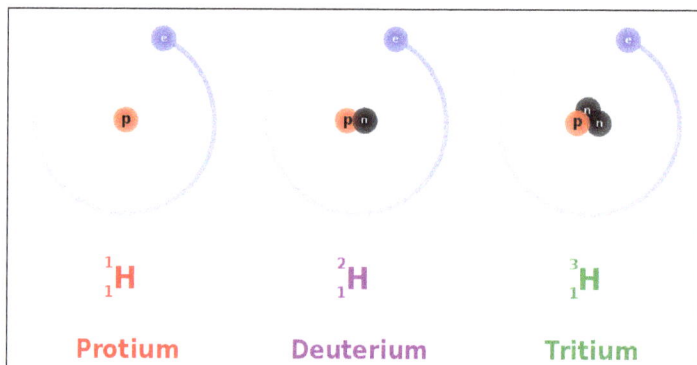

The three most stable isotopes of hydrogen: protium (A = 1), deuterium (A = 2), and tritium (A = 3).

List of Isotopes

Isotopes of Hydrogen	Number of Protons or atomic number	Number of Neutrons	Atomic Mass	Symbol	Natural Abundance (%)
Protium *or* Hydrogen-1	1	0	1.00782504	^1_1H or ^1H	> 99.98
Deuterium *or* Hydrogen-2	1	1	2.01410178	^2_1H or ^2H	0.0026 – 0.0184
Tritium *or* Hydrogen-3	1	2	3.0160492	^3_1H or ^3H	Very small trace amounts

- ( ) – Uncertainty (1σ) is given in concise form in parentheses after the corresponding last digits.

- Modes of decay:

  ○ n: Neutron emission.

- Bold for stable isotopes.

- Refers to that in water.

- Unless proton decay occurs.

- This and ^3He are the only stable nuclides with more protons than neutrons.

- Produced during Big Bang nucleosynthesis.

- Tank hydrogen has a^2H abundance as low as 3.2×10^{-5} (mole fraction).

- Produced during Big Bang nucleosynthesis, but not primordial, as all such atoms have since decayed to ^3He.

- Cosmogenic.

Hydrogen-1 (protium)

^{1}H (atomic mass 1.007825032241(94) u) is the most common hydrogen isotope with an abundance of more than 99.98%. Because the nucleus of this isotope consists of only a single proton, it is given the formal name *protium*.

The proton has never been observed to decay, and hydrogen-1 is therefore considered a stable isotope. Some grand unified theories proposed in the 1970s predict that proton decay can occur with a half-life between 10^{31} and 10^{36} years. If this prediction is found to be true, then hydrogen-1 (and indeed all nuclei now believed to be stable) are only observationally stable. To date, however, experiments have shown that the minimum proton half-life is in excess of 10^{34} years.

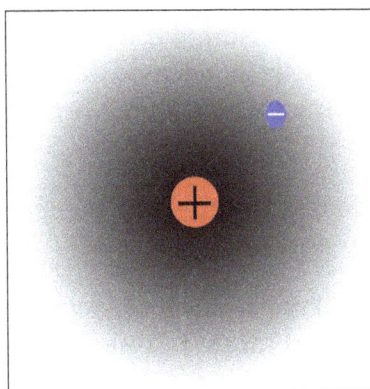

Protium, the most common isotope of hydrogen, consists of one proton and one electron. Unique among all stable isotopes, it has no neutrons.

Hydrogen-2 (Deuterium)

^2H (atomic mass 2.01410177811(12) u), the other stable hydrogen isotope, is known as deuterium and contains one proton and one neutron in its nucleus. The nucleus of deuterium is called a deuteron. Deuterium comprises 0.0026 – 0.0184% (by population, not by mass) of hydrogen samples on Earth, with the lower number tending to be found in samples of hydrogen gas and the higher enrichment (0.015% or 150 ppm) typical of ocean water. Deuterium on Earth has been enriched with respect to its initial concentration in the Big Bang and the outer solar system (about 27 ppm, by atom fraction) and its concentration in older parts of the Milky Way galaxy (about 23 ppm). Presumably the differential concentration of deuterium in the inner solar system is due to the lower volatility of deuterium gas and compounds, enriching deuterium fractions in comets and planets exposed to significant heat from the Sun over billions of years of solar system evolution.

Deuterium is not radioactive, and does not represent a significant toxicity hazard. Water enriched in molecules that include deuterium instead of protium is called heavy water. Deuterium and its compounds are used as a non-radioactive label in chemical experiments and in solvents for ^1H -NMR spectroscopy. Heavy water is used as a neutron moderator and coolant for nuclear reactors. Deuterium is also a potential fuel for commercial nuclear fusion.

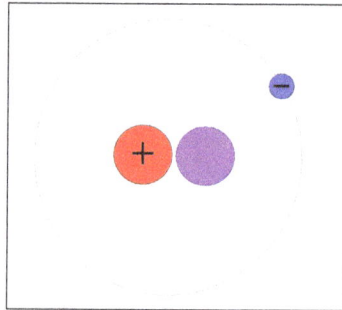

A deuterium atom contains one proton, one neutron, and one electron.

Hydrogen-3 (Tritium)

^3H (atomic mass 3.01604928199(23) u) is known as tritium and contains one proton and two neutrons in its nucleus. It is radioactive, decaying into helium-3 through β – decay with a half-life of 12.32 years. Trace amounts of tritium occur naturally because of the interaction of cosmic rays with atmospheric gases. Tritium has also been released during nuclear weapons tests. It is used in thermonuclear fusion weapons, as a tracer in isotope geochemistry, and specialized in self-powered lighting devices.

The most common method of producing tritium is by bombarding a natural isotope of lithium, lithium-6, with neutrons in a nuclear reactor.

Tritium was once used routinely in chemical and biological labeling experiments as a radiolabel, which has become less common in recent times. D-T nuclear fusion uses tritium as its main reactant, along with deuterium, liberating energy through the loss of mass when the two nuclei collide and fuse at high temperatures.

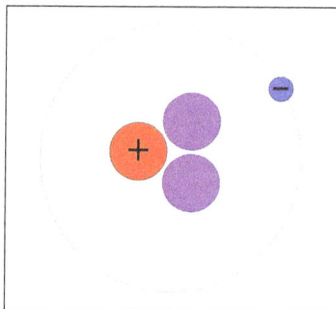

A tritium atom contains one proton, two neutrons, and one electron.

Hydrogen-4

^4H (atomic mass is 4.02643(11) u) contains one proton and three neutrons in its nucleus. It is a highly unstable isotope of hydrogen. It has been synthesised in the laboratory by bombarding tritium with fast-moving deuterium nuclei. In this experiment, the tritium nucleus captured a neutron from the fast-moving deuterium nucleus. The presence of the hydrogen-4 was deduced by detecting the emitted protons. It decays through neutron emission into hydrogen-3 (tritium) with a half-life of about 139 ± 10 yoctoseconds (or $(1.39 \pm 0.10) \times 10^{-22}$ seconds).

In the 1955 satirical novel The Mouse That Roared, the name quadium was given to the hydrogen-4 isotope that powered the Q-bomb that the Duchy of Grand Fenwick captured from the United States.

Hydrogen-5

^5H is a highly unstable isotope of hydrogen. The nucleus consists of a proton and four neutrons. It has been synthesised in the laboratory by bombarding tritium with fast-moving tritium nuclei. In this experiment, one tritium nucleus captures two neutrons from the other, becoming a nucleus with one proton and four neutrons. The remaining proton may be detected, and the existence of hydrogen-5 deduced. It decays through double neutron emission into hydrogen-3 (tritium) and has a half-life of at least 910 yoctoseconds $(9.1 \times 10^{-22}\ seconds)$.

Hydrogen-6

^6H decays either through triple neutron emission into hydrogen-3 (tritium) or quadruple neutron emission into hydrogen-2 (deuterium) and has a half-life of 290 yoctoseconds $(2.9 \times 10^{-22}\ seconds)$.

Hydrogen-7

^7H consists of a proton and six neutrons. It was first synthesised in 2003 by a group of Russian, Japanese and French scientists at RIKEN's Radioactive Isotope Beam Factory by bombarding hydrogen with helium-8 atoms. In the resulting reaction, all six of the helium-8's neutrons were donated to the hydrogen's nucleus. The two remaining protons were detected by the "RIKEN telescope", a device composed of several layers of sensors, positioned behind the target of the RI Beam cyclotron. Hydrogen-7 has a half life of 23 yoctoseconds $(2.3 \times 10^{-23}\ seconds)$.

Decay Chains

The majority of heavy hydrogen isotopes decay directly to ^3H, which then decays to the stable isotope ^3He. However, ^6H has occasionally been observed to decay directly to stable ^2H.

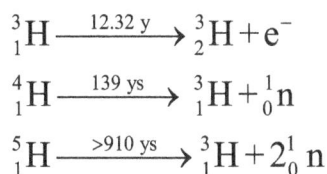

$$\ce{^3_1H ->[12.32\ y] ^3_2H + e^-}$$

$$\ce{^4_1H ->[139\ ys] ^3_1H + ^1_0n}$$

$$\ce{^5_1H ->[>910\ ys] ^3_1H + 2^1_0n}$$

$$^6_1\text{H} \xrightarrow{\;290\text{ ys}\;} {}^3_1\text{H} + 3^1_0\text{n}$$

$$^6_1\text{H} \xrightarrow{\;290\text{ ys}\;} {}^2_1\text{H} + 4^1_0\text{n}$$

$$^7_1\text{H} \xrightarrow{\;23\text{ ys}\;} {}^3_1\text{H} + 4^1_0\text{n}$$

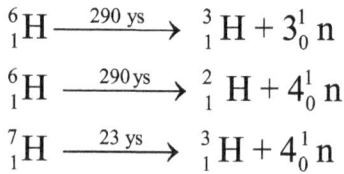

Note that the decay times are in yoctoseconds for all isotopes except ^3H, which is expressed in years.

HYDROGEN-LIKE ATOMS

A hydrogen-like ion is any atomic nucleus which has one electron and thus is isoelectronic with hydrogen. These ions carry the positive charge $e(Z-1)$ where Z is the atomic number of the atom. Examples of hydrogen-like ions are $\text{He}^+, \text{Li}^{2+}, \text{Be}^{3+}$ and B^{4+}. Because hydrogen-like ions are two-particle systems with an interaction depending only on the distance between the two particles, their (non-relativistic) Schrödinger equation can be solved in analytic form, as can the (relativistic) Dirac equation. The solutions are one-electron functions and are referred to as hydrogen-like atomic orbitals.

Other systems may also be referred to as "hydrogen-like atoms", such as muonium (an electron orbiting an antimuon), positronium (an electron and a positron), certain exotic atoms (formed with other particles), or Rydberg atoms (in which one electron is in such a high energy state that it sees the rest of the atom practically as a point charge).

Schrödinger Solution

In the solution to the Schrödinger equation, which is non-relativistic, hydrogen-like atomic orbitals are eigenfunctions of the one-electron angular momentum operator L and its z component L_z A hydrogen-like atomic orbital is uniquely identified by the values of the principal quantum number n, the angular momentum quantum number , and the magnetic quantum number . The energy eigenvalues do not depend on l or m, but solely on n. To these must be added the two-valued spin quantum number $m_s = \pm\frac{1}{2}$ setting the stage for the Aufbau principle. This principle restricts the allowed values of the four quantum numbers in electron configurations of more-electron atoms. In hydrogen-like atoms all degenerate orbitals of fixed n and l, m and s varying between certain values form an atomic shell.

The Schrödinger equation of atoms or atomic ions with more than one electron has not been solved analytically, because of the computational difficulty imposed by the Coulomb interaction between the electrons. Numerical methods must be applied in order to obtain (approximate) wavefunctions or other properties from quantum mechanical calculations. Due to the spherical symmetry (of the Hamiltonian), the total angular momentum J of an atom is a conserved quantity. Many numerical procedures start from products of atomic orbitals that are eigenfunctions of the one-electron operators L and L_z.

The radial parts of these atomic orbitals are sometimes numerical tables or are sometimes Slater orbitals. By angular momentum coupling many-electron eigenfunctions of J^2 (and possibly S^2) are constructed.

In quantum chemical calculations hydrogen-like atomic orbitals cannot serve as an expansion basis, because they are not complete. The non-square-integrable continuum (E > 0) states must be included to obtain a complete set, i.e., to span all of one-electron Hilbert space.

In the simplest model, the atomic orbitals of hydrogen-like ions are solutions to the Schrödinger equation in a spherically symmetric potential. In this case, the potential term is the potential given by Coulomb's law:

$$V(r) = -\frac{1}{4\pi\varepsilon_0}\frac{Ze^2}{r}$$

Where,

- $\varepsilon 0$ is the permittivity of the vacuum,

- Z is the atomic number (number of protons in the nucleus),

- e is the elementary charge (charge of an electron),

- r is the distance of the electron from the nucleus.

After writing the wave function as a product of functions:

$$\psi(r,\theta,\phi) = R_{nl}(r)Y_{lm}(\theta, \phi)$$

(in spherical coordinates), where Y_{lm} are spherical harmonics, we arrive at the following Schrödinger equation:

$$-\frac{\hbar^2}{2\mu}\left[\frac{1}{r^2}\frac{\partial}{\partial r}\left(r^2\frac{\partial R(r)}{\partial r}\right) - \frac{l(l+1)R(r)}{r^2}\right] + V(r)R(r) = ER(r),$$

where μ is, approximately, the mass of the electron (more accurately, it is the reduced mass of the system consisting of the electron and the nucleus), and $\hbar$ is the reduced Planck constant.

Different values of l give solutions with different angular momentum, where l (a non-negative integer) is the quantum number of the orbital angular momentum. The magnetic quantum number m (satisfying $-l \le m \le l$) is the (quantized) projection of the orbital angular momentum on the z-axis.

Non-relativistic Wavefunction and Energy

In addition to l and m, a third integer $n > 0$, emerges from the boundary conditions placed on R. The functions R and Y that solve the equations above depend on the values of these integers, called quantum numbers. It is customary to subscript the wave functions with the values of the quantum numbers they depend on. The final expression for the normalized wave function is:

$$\psi_{nlm} = R_{nl}(r)Y_{lm}(\theta,\phi)$$

$$R_{nl}(r) = \sqrt{\left(\frac{2Z}{na_\mu}\right)^3 \frac{(n-1-1)!}{2n\left[(n+1)!\right]}} e^{-Zr/na\mu} \left(\frac{2Zr}{na_\mu}\right)^l L_{n-l-1}^{2l+1}\left(\frac{2Zr}{na\mu}\right)$$

where:

- L_{n-l-1}^{2l+1} are the generalized Laguerre polynomials in the definition given here.

- $a\mu = \dfrac{4\pi\varepsilon_0\hbar^2}{\mu e^2} = \dfrac{\hbar_c}{\alpha\mu c^2} = \dfrac{m_e}{\mu} a_0$

  where α is the fine structure constant. Here, μ is the reduced mass of the nucleus-electron system, that is, $\mu = \dfrac{m_N m_e}{m_N + m_e}$ where m_N is the mass of the nucleus. Typically, the nucleus is much more massive than the electron, So $\mu \approx\sim m_e$. (But for positronium $\mu = m_e / 2$.)

- $E_n = -\left(\dfrac{Z^2\mu e^4}{32\mu^2\varepsilon_0^2\hbar^2}\right)\dfrac{1}{n^2} = -\left(\dfrac{Z^2\hbar^2}{2\mu a_\mu^2}\right)\dfrac{1}{n^2} = -\dfrac{\mu c^2 Z^2\alpha^2}{2n^2}$

- $Y_{lm}(\theta,\phi)$ function is a spherical harmonic

  parity due to angular wave function is $(-1)^l$.

Quantum Numbers

The quantum numbers n, l and m are integers and can have the following values:

$$n = 1, 2, 3, 4, \ldots$$
$$l = 0, 1, 2, \ldots, n-1$$
$$m = -l, -l+1, \ldots, 0, \ldots, l-1, l.$$

Angular Momentum

Each atomic orbital is associated with an angular momentum L. It is a vector operator, and the eigenvalues of its square $L^2 \equiv L_x^2 + L_y^2 + L_z^2$ are given by:

$$L^2 Y_{lm} = \hbar^2 l(l+1)Y_{lm}$$

The projection of this vector onto an arbitrary direction is quantized. If the arbitrary direction is called z, the quantization is given by:

$$L_z Y_{lm} = \hbar m Y_{lm},$$

where m is restricted as described above. Note that L_2 and L_z commute and have a common eigenstate, which is in accordance with Heisenberg's uncertainty principle. Since L_x and L_y do not commute with L_z, it is not possible to find a state that is an eigenstate of all three components simultaneously. Hence the values of the x and y components are not sharp, but are given by a probability function of finite width. The fact that the x and y components are not well-determined, implies that the direction of the angular momentum vector is not well determined either, although its component along the z-axis is sharp.

These relations do not give the total angular momentum of the electron. For that, electron spin must be included.

This quantization of angular momentum closely parallels that proposed by Niels Bohr in 1913, with no knowledge of wavefunctions.

Including Spin-orbit Interaction

In 1928 in England Paul Dirac found an equation that was fully compatible with Special Relativity. The equation was solved for hydrogen-like atoms the same year (assuming a simple Coulomb potential around a point charge) by the German Walter Gordon. Instead of a single (possibly complex) function as in the Schrödinger equation, one must find four complex functions that make up a bispinor. The first and second functions (or components of the spinor) correspond (in the usual basis) to spin "up" and spin "down" states, as do the third and fourth components.

The terms "spin up" and "spin down" are relative to a chosen direction, conventionally the z direction. An electron may be in a superposition of spin up and spin down, which corresponds to the spin axis pointing in some other direction. The spin state may depend on location.

An electron in the vicinity of a nucleus necessarily has non-zero amplitudes for the third and fourth components. Far from the nucleus these may be small, but near the nucleus they become large.

The eigenfunctions of the Hamiltonian, which means functions with a definite energy (and which therefore do not evolve except for a phase shift), have energies characterized not by the quantum number n only (as for the Schrödinger equation), but by n and a quantum number j, the total angular momentum quantum number. The quantum number j determines the sum of the squares of the three angular momenta to be $j(j+1)$ (times $\hbar^2$,). These angular momenta include both orbital angular momentum (having to do with the angular dependence of ψ) and spin angular momentum (having to do with the spin state). The splitting of the energies of states of the same principal quantum number n due to differences in j is called fine structure. The total angular momentum quantum number j ranges from $1/2$ to $n-1/2$.

The orbitals for a given state can be written using two radial functions and two angle functions. The radial functions depend on both the principal quantum number n and an integer k, defined as:

$$k = \begin{cases} -j-\dfrac{1}{2} & \text{if } j = \ell + \dfrac{1}{2} \\ \\ j+\dfrac{1}{2} & \text{if } j = \ell - \dfrac{1}{2} \end{cases}$$

where ℓ is the azimuthal quantum number that ranges from 0 to $n-1$. The angle functions depend on k and on a quantum number m which ranges from $-j$ to j by steps of 1. The states are labeled using the letters S, P, D, F et cetera to stand for states with ℓ equal to 0, 1, 2, 3 et cetera, with a subscript giving j. For instance, the states for $n = 4$ are given in the following table (these would be prefaced by n, for example $4S_{1/2}$):

	m = -7/2	m = -5/2	m = -3/2	m = -1/2	m = 1/2	m = 3/2	m = 5/2	m = 7/2
$k = 3, \ell = 3$		$F_{5/2}$	$F_{5/2}$	$F_{5/2}$	$F_{5/2}$	$F_{5/2}$	$F_{5/2}$	
$k = 2, \ell = 2$			$D_{3/2}$	$D_{3/2}$	$_{3/2}$	$D_{3/2}$		
$k = 1, \ell = 1$				$P_{1/2}$	$P_{1/2}$			
$k = 0$								
$k = -1, \ell = 0$				$S_{1/2}$	$S_{1/2}$			
$k = -2, \ell = 1$			$P_{3/2}$	$P_{3/2}$	$P_{3/2}$	$P_{3/2}$		
$k = -3, \ell = 2$		$D_{5/2}$	$D_{5/2}$	$D_{5/2}$	$D_{5/2}$	$D_{5/2}$	$D_{5/2}$	
$k = -4, \ell = 3$	$F_{7/2}$	$F_{7/2}$	$F_{7/2}$	$F_{7/2}$	$F_{7/2}$	$F_{7/2}$	$F_{7/2}$	$F_{7/2}$

These can be additionally labeled with a subscript giving m. There are $2n^2$ states with principal quantum number n, $4j+2$ of them with any allowed j except the highest $(j = n-1/2)$ for which there are only $2j+1$. Since the orbitals having given values of n and j have the same energy according to the Dirac equation, they form a basis for the space of functions having that energy.

The energy, as a function of n and |k| (equal to $j+1/2$), is:

$$E_{nj} = \mu c^2 \left(1 + \left[\frac{Z\alpha}{n - |k| + \sqrt{k^2 - Z^2\alpha^2}}\right]^2\right)^{-1/2}$$

$$\approx \mu c^2 \left\{1 - \frac{Z^2\alpha^2}{2n^2}\left[1 + \frac{Z^2\alpha^2}{n}\left(\frac{1}{|k|} - \frac{3}{4n}\right)\right]\right\}$$

(The energy of course depends on the zero-point used.) Note that if Z were able to be more than 137 (higher than any known element) then we would have a negative value inside the square root for the $S_{1/2}$ and $P_{1/2}$ orbitals, which means they would not exist. The Schrödinger solution corresponds to replacing the inner bracket in the second expression by 1. The accuracy of the energy difference between the lowest two hydrogen states calculated from the Schrödinger solution is about 9 ppm (90 μeV too low, out of around 10 eV), whereas the accuracy of the Dirac equation for the same energy difference is about 3 ppm (too high). The Schrödinger solution always puts the states at slightly higher energies than the more accurate Dirac equation. The Dirac equation gives some levels of hydrogen quite accurately (for instance the $4P_{1/2}$ state is given an energy only about 2×10^{-10} eV too high), others less so (for instance, the $2S_{1/2}$ level is about 4×10^{-6} eV too low). The modifications of the energy due to using the Dirac equation rather than the Schrödinger solution is of the order of α^2, and for this reason α is called the fine structure constant.

The solution to the Dirac equation for quantum numbers n, k, and m, is:

$$\Psi = \begin{pmatrix} g_{n,k}(r)r^{-1}\,\Omega_{k,m}(\theta,\phi) \\ i\,f_{n,k}(r)^{r-1}\,\Omega_{-k,m}(\theta,\phi) \end{pmatrix} = \begin{pmatrix} g_{n,k}(r)r^{-1}\sqrt{\left(k+\dfrac{1}{2}-m\right)/(2k+1)}\;Y_{k,m-1/2}(\theta,\phi) \\[2ex] -g_{n,k}(r)r^{-1}\operatorname{sgn}k\sqrt{\left(k+\dfrac{1}{2}+m\right)/(2k+1)}\;Y_{k,m+1/2}(\theta,\phi) \\[2ex] i\,f_{n,k}(r)r^{-1}\sqrt{\left(-k+\dfrac{1}{2}-m\right)/(-2k+1)}\;Y_{-k,m-1/2}(\theta,\phi) \\[2ex] -i\,f_{n,k}(r)r^{-1}\operatorname{sgn}k\sqrt{\left(-k+\dfrac{1}{2}-m\right)/(-2k+1)}\;Y_{-k,m+1/2}(\theta,\phi) \end{pmatrix}$$

where the Ωs are columns of the two spherical harmonics functions shown to the right. $Y_{a,b}(\theta,\phi)$ signifies a spherical harmonic function:

$$Y_{a,b}(\theta,\phi) = \begin{cases} (-1)^b\sqrt{\dfrac{2a+1}{4\pi}\dfrac{(a-b)!}{(a+b)!}}\,P_a^b(\cos\theta)e^{ib\phi} & \text{if } a > 0 \\[2ex] Y_{-a-1,b}(\theta,\phi) & \text{if } a < 0 \end{cases}$$

in which P_a^b is an associated Legendre polynomial. (Note that the definition of Ω may involve a spherical harmonic that doesn't exist, like $Y_{0,1}$, but the coefficient on it will be zero.)

Here is the behavior of some of these angular functions. The normalization factor is left out to simplify the expressions.

$$\Omega_{-1,-1/2}\;\alpha\begin{pmatrix} 0 \\ 1 \end{pmatrix}$$

$$\Omega_{-1,1/2}\;\alpha\begin{pmatrix} 1 \\ 0 \end{pmatrix}$$

$$\Omega_{1,-1/2}\;\alpha\begin{pmatrix} (x-iy)/r \\ z/r \end{pmatrix}$$

$$\Omega_{1,1/2}\;\alpha\begin{pmatrix} z/r \\ (x+iy)/r \end{pmatrix}$$

From these we see that in the $S_{1/2}$ orbital $(k = -1)$, the top two components of Ψ have zero orbital angular momentum like Schrödinger S orbitals, but the bottom two components are orbitals like the Schrödinger P orbitals. In the $P_{1/2}$ solution $(k = 1)$, the situation is reversed. In both cases, the spin of each component compensates for its orbital angular momentum around the z axis to give the right value for the total angular momentum around the z axis.

The two Ω spinors obey the relationship:

$$\Omega_{k,m} = \begin{pmatrix} z/r & (x-iy)/r \\ (x+iy) & -z/r \end{pmatrix} \Omega_{-k,m}$$

To write the functions $g_{n,k}(r)$ and $f_{n,k}(r)$ let us define a scaled radius ρ:

$$\rho \equiv 2Cr$$

With:

$$C = \frac{\sqrt{\mu^2 c^4 - E^2}}{\hbar c}$$

where E is the energy $\left(E_{n,j}\right)$ given above. We also define γ as:

$$\gamma \equiv \sqrt{k^2 - Z^2 \alpha^2}$$

When $k = -n$ (which corresponds to the highest j possible for a given n, such as $1S_{1/2}$, $2P_{3/2}$, $3D_{5/2}$...), then $g_{n,k}(r)$ and $f_{n,k}(r)$ are:

$$g_{n,-n}(r) = A(n+\gamma)\rho^\gamma e^{-\rho/2}$$
$$f_{n,-n}(r) = AZ\alpha\rho^\gamma e^{-\rho/2}$$

where A is a normalization constant involving the Gamma function:

$$A = \frac{1}{\sqrt{2n(n+\gamma)}} \sqrt{\frac{C}{\gamma \Gamma(2\gamma)}}$$

Notice that because of the factor $Z\alpha$, f(r) is small compared to g(r). Also notice that in this case, the energy is given by:

$$E_{n,n-1/2} = \frac{\gamma}{n}\mu c^2 = \sqrt{1 - \frac{Z^2 \alpha^2}{n^2}}\mu c^2$$

and the radial decay constant C by:

$$C = \frac{Z\alpha}{n}\frac{\mu c^2}{\hbar c}.$$

In the general case (when k is not $-n$), $g_{n,k}(r)$ and $f_{n,k}(r)$ are based on two generalized Laguerre polynomials of order $n-|k|-1$ and $n-|k|$:

$$g_{n,k}(r) = A\rho^\gamma e^{-\rho/2}\left(Z\alpha\rho L_{n-|k|-1}^{2\gamma+1}(\rho) + (\gamma-k)\frac{\gamma\mu c^2 - kE}{\hbar c C} L_{n-|k|}^{2\gamma-1}(\rho)\right)$$

$$f_{n,k}(r) = A\rho^{\gamma}e^{-\rho/2}(\gamma - k)\rho L_{n-|k|-1}^{2\gamma+1}(\rho) + Z\alpha\frac{\gamma\mu c^2 - kE}{\hbar cC}L_{n-|k|}^{2\gamma-1}(\rho)$$

with A now defined as:

$$A = \frac{1}{\sqrt{2k(k-\gamma)}}\sqrt{\frac{C}{n-|k|+\gamma}\frac{(n-|k|-1)!}{\Gamma(n-|k|+2\gamma+1)}\frac{1}{2}\left(\left(\frac{Ek}{\gamma\mu c^2}\right)^2 + \frac{Ek}{\gamma\mu c^2}\right)}$$

Again f is small compared to g (except at very small r) because when k is positive the first terms dominate, and α is big compared to $\gamma - k$, whereas when k is negative the second terms dominate and α is small compared to $\gamma - k$. Note that the dominant term is quite similar to corresponding the Schrödinger solution – the upper index on the Laguerre polynomial is slightly less ($2\gamma + 1$ or $2\gamma - 1$ rather than $2\ell + 1$, which is the nearest integer), as is the power of ρ (γ or $\gamma - 1$ instead of ℓ, the nearest integer). The exponential decay is slightly faster than in the Schrödinger solution.

The normalization factor makes the integral over all space of the square of the absolute value equal to 1.

1S Orbital

Here is the $1S_{1/2}$ orbital, spin up, without normalization:

$$\Psi\alpha\begin{pmatrix}(1+\gamma)r^{\gamma-1}e^{-Cr}\\0\\iZ\alpha r^{\gamma-1}e^{-Cr}z/r\\iZ\alpha r^{\gamma-1}e^{-Cr}(x+iy)/r\end{pmatrix}$$

Note that γ is a little less than 1, so the top function is similar to an exponentially decreasing function of r except that at very small r it theoretically goes to infinity. But the value of the $r^{\gamma-1}$ only surpasses 10 at a value of r smaller than $10^{1/(\gamma-1)}$, which is a very small number (much less than the radius of a proton) unless Z is very large.

The $1S_{1/2}$ orbital, spin down, without normalization, comes out as:

$$\Psi\alpha\begin{pmatrix}0\\(1+\gamma)r^{\gamma-1}e^{-Cr}\\iZ\alpha r^{\gamma-1}e^{-Cr}(x-iy)/r\\-iZ\alpha r^{\gamma-1}e^{-Cr}z/r\end{pmatrix}$$

We can mix these in order to obtain orbitals with the spin oriented in some other direction, such as:

$$\Psi \; \alpha \begin{pmatrix} (1+\gamma)r^{\gamma-1}e^{-Cr} \\ (1+\gamma)r^{\gamma-1}e^{-Cr} \\ iZ\alpha\, r^{\gamma-1}e^{-Cr}(x-iy+z)/r \\ -iZ\alpha\, r^{\gamma-1}e^{-Cr}(x+iy-z)/r \end{pmatrix}$$

which corresponds to the spin and angular momentum axis pointing in the x direction. Adding i times the "down" spin to the "up" spin gives an orbital oriented in the y direction.

2P$_{1/2}$ and 2S$_{1/2}$ orbitals

To give another example, the 2P$_{1/2}$ orbital, spin up, is proportional to:

$$\Psi \; \alpha \begin{pmatrix} \rho^{\gamma-1}e^{-\rho/2}\left(Z\alpha\rho+(\gamma-1)\dfrac{\gamma\mu c^2 - E}{\hbar cC}(-\rho+2\gamma)\right)z/r \\[2ex] \rho^{\gamma-1}e^{-\rho/2}\left(Z\alpha\rho+(\gamma-1)\dfrac{\gamma\mu c^2 - E}{\hbar cC}(-\rho+2\gamma)\right)(x+iy)/r \\[2ex] i\rho^{\gamma-1}e^{-\rho/2}\left((\gamma-1)\rho+Z\alpha\dfrac{\gamma\mu c^2 - E}{\hbar cC}(-\rho+2\gamma)\right) \\[2ex] 0 \end{pmatrix}$$

(Remember that $\rho = 2rC$. C is about half what it is for the 1S orbital, but γ is still the same.)

Notice that when ρ is small compared to α (or r is small compared to $\hbar c/(\mu c^2)$ the "S" type orbital dominates (the third component of the bispinor).

For the 2S$_{1/2}$ spin up orbital, we have:

$$\Psi \; \alpha \begin{pmatrix} \rho^{\gamma-1}e^{-\rho/2}\left(Z\alpha\rho+(\gamma+1)\dfrac{\gamma\mu c^2 + E}{\hbar cC}(-\rho+2\gamma)\right) \\[2ex] 0 \\[2ex] i\rho^{\gamma-1}e^{-\rho/2}\left((\gamma+1)\rho+Z\alpha\dfrac{\gamma\mu c^2 + E}{\hbar cC}(-\rho+2\gamma)\right)z/r \\[2ex] i\rho^{\gamma-1}e^{-\rho/2}\left((\gamma+1)\rho+Z\alpha\dfrac{\gamma\mu c^2 + E}{\hbar cC}(-\rho+2\gamma)\right)(x+iy)/r \end{pmatrix}$$

Now the first component is S-like and there is a radius near $\rho = 2$ where it goes to zero, whereas the bottom two-component part is P-like.

Negative-energy Solutions

In addition to bound states, in which the energy is less than that of an electron infinitely separated from the nucleus, there are solutions to the Dirac equation at higher energy, corresponding to an unbound electron interacting with the nucleus. These solutions are not normalizable, but solutions can be found which tend toward zero as r goes to infinity (which is not possible when $|E| < \mu c^2$ except at the above-mentioned bound-state values of E). There are similar solutions with $E < -\mu c^2$. These negative-energy solutions are just like positive-energy solutions having the opposite energy but for a case in which the nucleus repels the electron instead of attracting it, except that the solutions for the top two components switch places with those for the bottom two.

Negative-energy solutions to Dirac's equation exist even in the absence of a Coulomb force exerted by a nucleus. Dirac hypothesized that we can consider almost all of these states to be already filled. If one of these negative-energy states is not filled, this manifests itself as though there is an electron which is repelled by a positively-charged nucleus. This prompted Dirac to hypothesize the existence of positively-charged electrons, and his prediction was confirmed with the discovery of the positron.

Beyond Gordon's Solution to the Dirac Equation

The Dirac equation with a simple Coulomb potential generated by a point-like non-magnetic nucleus was not the last word, and its predictions differ from experimental results as mentioned earlier. More accurate results include the Lamb shift (radiative corrections arising from quantum electrodynamics) and hyperfine structure.

MULTI ELECTRON ATOMS

Electrons with more than one atom, such as Helium (He), and Nitrogen (N), are referred to as multi-electron atoms. Hydrogen is the only atom in the periodic table that has one electron in the orbitals under ground state. We will learn how additional electrons behave and affect a certain atom.

First, electrons repel against each other. Particles with the same charge repel each other, while oppositely charged particles attract each other. For example, a proton, which is positively charged, is attracted to electrons, which are negatively charged. However, if we put two electrons together or two protons together, they will repel one another. Since neutrons lack a charge, they will neither repel nor attract protons or electrons.

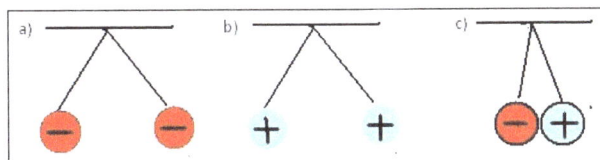

The two electrons are placed together and repel each other because of the same charge. b) The two protons are repelling each other for the same reason. c) When oppositely charged particles, an electron and a proton, are placed together, they attract each other.

Protons and neutrons are located in an atom's nucleus. Electrons float around the atom in energy levels. Energy levels consist of orbitals and sub-orbitals. The lower the energy level the electron is located at, the closer it is to nucleus. As we go down and to the right of the periodic table, the number of electrons that an element has increases. Since there are more electrons, the atom experiences greater repulsion and electrons will tend to stay as far away from each other as possible. Our main focus is what effects take place when more electrons surround the nucleus.

Shielding (Screening)

With more protons in the nucleus, the attractive force for electrons to the nucleus is stronger. Thus, the orbital energy becomes more negative (less energy). Orbital energy also depends on the type of l orbital an electron is located in. The lower the number of l, the closer it is to the nucleus. For example, $l = 0$ is the s orbital. S orbitals are closer to the nucleus than the p orbitals $(l = 1)$ that are closer to the nucleus than the d orbitals $(l = 2)$ that are closer to the f orbitals $(l = 3)$.

More electrons create the shielding or screening effect. Shielding or screening means the electrons closer to the nucleus block the outer valence electrons from getting close to the nucleus. Imagine being in a crowded auditorium in a concert. The enthusiastic fans are going to surround the auditorium, trying to get as close to the celebrity on the stage as possible. They are going to prevent people in the back from seeing the celebrity or even the stage. This is the shielding or screening effect. The stage is the nucleus and the celebrity is the protons. The fans are the electrons. Electrons closest to the nucleus will try to be as close to the nucleus as possible. The outer/valence electrons that are farther away from the nucleus will be shielded by the inner electrons. Therefore, the inner electrons prevent the outer electrons from forming a strong attraction to the nucleus. The degree to which the electrons are being screened by inner electrons can be shown by ns<np<nd<nf where n is the energy level. The inner electrons will be attracted to the nucleus much more than the outer electrons. Thus, the attractive forces of the valence electrons to the nucleus are reduced due to the shielding effects. That is why it is easier to remove valence electrons than the inner electrons. It also reduces the nuclear charge of an atom.

Penetration

Penetration is the ability of an electron to get close to the nucleus. The penetration of ns > np > nd > nf. Thus, the closer the electron is to the nucleus, the higher the penetration. Electrons with higher penetration will shield outer electrons from the nucleus more effectively. The s orbital is closer to the nucleus than the p orbital. Thus, electrons in the s orbital have a higher penetration than electrons in the p orbital. That is why the s orbital electrons shield the electrons from the p orbitals. Electrons with higher penetration are closer to the nucleus than electrons with lower penetration. Electrons with lower penetration are being shielded from the nucleus more.

Radial Probability Distribution

Radial probability distribution is a type of probability to find where an electron is mostly likely going to be in an atom. The higher the penetration, the higher probability of finding an electron near the nucleus. As shown by the graphs, electrons of the s orbital are found closer to the nucleus than the p orbital electrons. Likewise, the lower the energy level an electron is located at, the higher chance it has of being found near the nucleus. The smaller the energy level (n) and the orbital angular momentum quantum number (l) of an electron is, the more likely it will be near the nucleus. As electrons get to higher and higher energy levels, the harder it is to locate it because the radius of the sphere is greater. Thus, the probability of locating an electron will be more difficult. Radial probability distribution can be found by multiplying $4\pi r^2$, the area of a sphere with a radius of r and $R^2(r)$.

$$\text{Radial Probability Distribution} = 4\pi r^2 \; X \; R^2(r)$$

By using the radial probability distribution equation, we can get a better understanding about an electron's behavior.

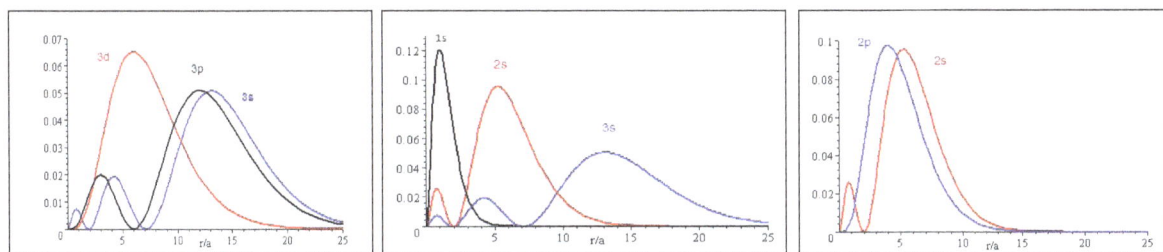

The lower the energy level, the higher the probability of finding the electron close to the nucleus. Also, the lower momentum quantum number gets, the closer it is to the nucleus.

Effective Nuclear Charge (Z_{eff})

Effective nuclear charge (Z_{eff}) is the positive nuclear charge that is experienced by an electron. The more electrons shielding an electron has, the lower the Z_{eff} value. Since it is farther away from the nucleus, the force is weaker. Using the analogy of the stage, imagine you are in the back. With all the noise and shielding from the inner people, you can barely see the stage or even hear the celebrity. Likewise, the outer electrons will have a lower value of Z_{eff}. Thus, the lower the Z_{eff} value, the less attracted the electron is to the nucleus. So, it is not held tightly by the nucleus. As a result, the electron has to be in a higher energy level orbital because it farther away from the nucleus. Therefore, electrons in the lower energy s orbital with higher penetration are less shielded by other electrons and experience a higher Z_{eff} than p orbital electrons. That is why orbitals have different energy. 1s electrons have less energy than 2s electrons. 1s electrons shield the 2s electrons from "seeing" the nucleus because it has a higher penetration.

How do we calculate Ze_{eff}?

$$Z_{eff} = Z - S$$

Z_{eff} is effective nuclear charge

Z is the atomic number

S is the number of inner energy level/ core electrons

For example:

Z_{eff} of Li $= 3 - 2 = 1$

How do effective nuclear charge and energy level affect the orbital energy?

$$E_n \propto \frac{-Z^2_{eff}}{n^2}$$

En is the orbital energy
Zeff is the effective nuclear charge
n is the energy level

Z_{eff} is directly proportional to En and is inversely proportional to n. As Z_{eff} increases, En increases. As n increases, En decreases.

Remember the form for the Hamiltonian for the hydrogen-like atom:

$$\widehat{H} = \hat{T} + \hat{V} = -\frac{1}{2}\nabla^2 - \frac{Z}{R},$$

in atomic units $(m_e = \hbar = 1, e = -1)$, where R is the distance between the nucleus and the electron, and Z is the atomic number of the nucleus. Similar for a system of n electrons the Hamiltonian is,

$$\widehat{H} = -\frac{1}{2}\sum_{i=1}^{n}\nabla_i^2 - \sum_{i=1}^{n}\frac{Z}{R_i} + \frac{1}{2}\sum_{i,j=1}^{n}{}'\frac{1}{r_{ij}},$$

where the first term is the kinetic energy operator for each electron, the second term is due to the attraction between the electron and the nucleus, and the last term accounts for the repulsion due to electron-electron interactions. The factor $\frac{1}{2}$ in front of the double sum prevents counting the electronelectron interactions twice, and the prime excludes the $i = j$ terms. If we compare with the Hamiltonian for the hydrogen-like atom we realize that we can express the many-electron Hamiltonian as,

$$\widehat{H} = \sum_{i=1}^{n}\left[-\frac{1}{2}\nabla_i^2 - \frac{Z}{R_i}\right] + \frac{1}{2}\sum_{i,j=1}^{n}{}'\frac{1}{r_{ij}}$$

$$= \sum_{i=1}^{n}\hat{h}(i) + \frac{1}{2}\sum_{i,j=1}^{n}{}'\frac{1}{r_{ij}},$$

where $\hat{h}(i)$ is the hydrogen-like Hamiltonian for the i'th electron. Lets consider the simplest system having more than one electron which is the helium atom.

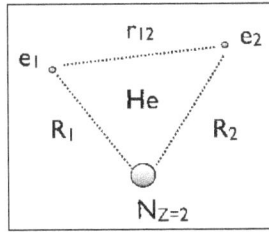

For He the Hamiltonian reads,

$$\widehat{H} = -\frac{1}{2}\nabla_1^2 - \frac{1}{2}\nabla_2^2 - \frac{2}{R_1} - \frac{2}{R_2} + \frac{1}{r_{12}}$$

$$= \hat{h}(1) + \hat{h}(2)\frac{1}{r_{12}}$$

We already know how to solve for $\hat{h}(1)$ and $\hat{h}(2)$, but what about the last term? Since the term depends on both the position of electron 1 and 2 we cannot separate their motion. This means that we cannot use the trick of separating variables which we have used so far and we cannot obtain an analytic solution to the Schrördinger equation.

The Independent Electron Approximation

The simplest approximation is simply to ignore the electron-electron repulsion. Obviously this is a very bad approximation but it allows us to separate variables and thereby illustrate some important physics of a many-electron system. Since we neglect the interactions between the electron s we treat the electrons as independent of each other, hence the name "independent electron approximation". The approximate Hamiltonian is then:

$$\widehat{H}_{approx.} = \hat{h}(1) + \hat{h}(2),$$

where each of the individual hydrogen-like Hamiltonians obey a one-electron Schrödinger equation,

$$\hat{h}(1)\phi_i(1) = \varepsilon_i\phi_i(1),$$

where ϕ_i is the atomic orbital and ε_i is the orbital energy. This mean than an orbital is simply a one-electron wavefunction. We already know the solution to these one-electron Schrödinger equation from our treatment of the hydrogen atom, i.e.,

$$\phi_i \in \{1s, 2s, 2p, \ldots\}, \varepsilon_i = -\frac{1}{2}\frac{Z^2}{n^2}.$$

Previously, we have shown that if we can separate the Hamiltonian into independent terms we can write the total wavefunction as product of eigenfunctions of the individual terms. Therefore, the wavefunction for He in the independent electron approximation becomes,

$$\psi(He) = \phi_i(1)\phi_j(2)$$

Now, lets show that this is wavefunction is an eigenfunction of the approximate Hamiltonian,

$$\widehat{H}_{approx} \psi(He) = \left(\hat{h}(1) + \hat{h}(2)\phi_i(1)\phi_j(2)\right)$$
$$= \hat{h}(1)\phi_i(1)\phi_j(2) + \hat{h}(2)\phi_i(1)\phi_j(2),$$

where $\hat{h}(1)$ and $\hat{h}(2)$ only works on variables of electron 1 and electron 2, respectively. This gives us,

$$\widehat{H}_{approx} \psi(He) = \hat{h}(1)\phi_i(1)\phi_j(2) + \phi_i(1)\hat{h}(2)\phi_j(2)$$
$$= \varepsilon_i\phi_i(1)\phi_j(2) + \phi_i(1)\varepsilon_j\phi_j(2)$$
$$= \left(\varepsilon_i + \varepsilon_j\right)\phi_i(1)\phi_j(2)$$
$$= \left(\varepsilon_i + \varepsilon_j\right)\psi(He)$$

and we see that the total energy is the sum of the individual orbital energies.

Electron Exchange Symmetry

We can now write the ground state of the He atom in the independent electron approximation as 1s(1)1s(2), for which the electron configuration is 1s². Now lets instead consider the first excited state of He. What would we expect? There are only two different ways of arranging the electrons in a simple product,

$$\psi(1, 2) = 1s(1)2s(2), \quad or, \quad \psi(2, 1) = 2s(1)1s(2),$$

which both have the same energy but describes different physics, e.g. electron distribution. This can be illustrated by considering the electron distribution,

$$\left|\psi(1,2)\right|^2 = \frac{8}{\pi}\exp(-4r_1)\frac{1}{\pi}\left(1 - 2r_2 + r_2^2\right)\exp(-2r_2),$$

And,

$$\left|\psi(2,1)\right|^2 = \frac{8}{\pi}\exp(-4r_2)\frac{1}{\pi}\left(1 - 2r_2 + r_1^2\right)\exp(-2r_2).$$

Therefore, we see that:

$$\left|\psi(1, 2)\right|^2 \neq \left|\psi(2, 1)\right|^2.$$

The two wavefunctions therefore differ by exchange of electron indices, something that is wrong. While in classical mechanics it is always possible to distinguish between identical particles, however, this is not the case in quantum mechanics due to the uncertainty principle. Therefore, in quantum mechanism identical particles are indistinguishable and the probability has to be invariant under exchange of indices. This implies that the wavefunction it self most by either symmetric or antisymmetric under the exchange. Let $\hat{P}_{ij}$ be an operator that when acting on a function interchange the indices i and j such that:

$$\hat{P}_{ij}\,\psi\,(ij) = \pm\psi\,(ij).$$

For He we can construct the following symmetric and antisymmetric wavefunction from linear combinations of the original functions:

$$\psi_s = \frac{1}{\sqrt{2}}\Big[1s(1)2s(2)+2s(1)1s(2)\Big],$$

And

$$\psi_s = \frac{1}{\sqrt{2}}\Big[1s(1)2s(2)-2s(1)1s(2)\Big].$$

If we now operate on these wave functions with $\hat{P}_{12}$ we see that:

$$\hat{P}_{12}\psi s = \frac{1}{\sqrt{2}}\Big[1s(2)2s(1)+2s(2)1s(1)\Big] = +\psi s$$

$$\hat{P}_{12}\psi s = \frac{1}{\sqrt{2}}\Big[1s(2)2s(1)-2s(2)1s(1)\Big] = +\psi s.$$

If we now plot the propability density of these new functions we will see that the antisymmetric function has a depletion of density as the distance goes to zero, which is called a Fermi hole. Similar the symmetry function has a increase in density near the region called a Fermi heap.

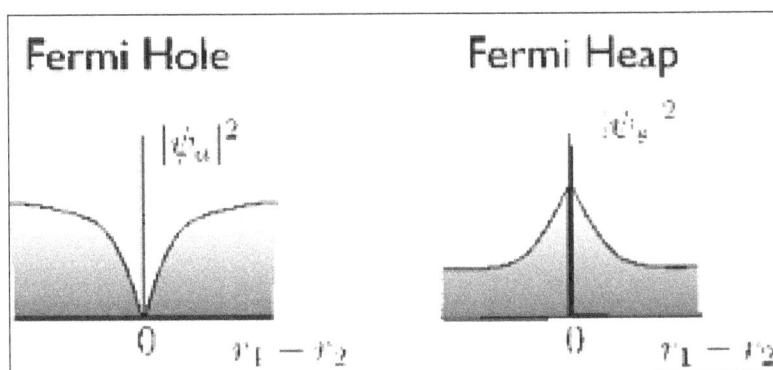

Stern-Gerlach Experiment

Up to this point, we have treated the particles as structureless with a mass and a charge which can be described by a wavefunction specified by the spatial coordinates, x, y, z, as variables. However, our empirical evidence points to the need for attributing an intrinsic angular momentum to the particles as well. The most direct evidence comes from the work of Stern and Gerlach in their now famous experiment from 1922, although it was not realized at that time. In the experiment they pass an unexcited beam of silver atoms though an inhomogeneous magnetic field.

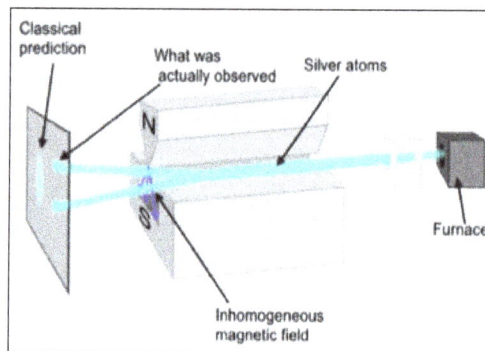

If a particle have a magnetic moment μ (we have seen that the angular momentum of a charged species gives rise to a magnetic moment $\mu = -\beta eL$) they will be deflected due to a force, $F \sim \mu b \nabla B$. Thus, particles with different moments will be deflected differing amounts by the magnetic field. If the particles are classical, "spinning" particles, then the distribution of their spin angular momentum vectors is taken to be truly random and each particle would be deflected up or down by a different amount, producing an even distribution on the screen of a detector. Consider the electronic configuration of silver, Ag = [Kr]4d105s1, what would you expect classical for the moment and what would you expect to see in the experiment?

Thus, the magnitude magnetic moment of the electron is fixed, and the direction it points is quantized and can take on one of two values. Since any component of L has $2l + 1$ eigenvalues we would expect that the magnetic moment to have $2l + 1$ eigenvalues. However, experimentally only two distinct traces where seen (although, nothing was seen due to very thin layer deposited until Stern's breath full of sulfur from cheap cigars developed AgS which is "jet" black). The solution to this comes from the postulate by Goudsmit and Uhlenbeck in 1925 that the electron have an intrinsic angular moment now called spin with an angular magnitude of $1/2\hbar$. They postulated the existents of the spin in order to explain the fine structure of atomic spectra. However, it was not until 1928 when Dirac combined Einsteins relativity with quantum mechanics that the treatment of electron spin arised naturally. This is well beyond the scope of this class.

Spin and the Pauli Exclusion Principle

The spin aungular momentum is denoted by $\hat{S}$ and similar to the orbital angular momentum they obey the following eigenvalue equations:

$$\hat{S}^2 \left| sm_s \right\rangle = \hbar^2 \left(s + 1 \right) \left| sm_s \right\rangle, \hat{S}_z \left| sm_s \right\rangle = \hbar m_s \left| sm_s \right\rangle$$

where the allowed values of $m_s = -s, -s + 1, \cdots, s - 1, s$. Electrons have a spin of $s = \dfrac{1}{2}$ which gives two values for $m_s = \pm \dfrac{1}{2}\hbar$. The eigenstate corresponding to $\left| \dfrac{1}{2}, \dfrac{1}{2} \right\rangle$ is traditional called α or spin-up and $\left| \dfrac{1}{2}, -\dfrac{1}{2} \right\rangle$ is called β or spin-down. We will not express then spin function in explicit form but rather keep the symbolic representation α and β. Therefore, the wavefunction $\psi = \phi(1)\alpha(1)$ refer to an electron in space orbital $\phi(1)$ with spin α , and is called a spin-orbital. Since the electrons are indistinguishable we require that the wavefunction is either symmetric or antisymmetric with respect to interchange of electron space and spin indices. Lets consider the possible combination of space and spin wavefunctions for the ground state of He:

$$1s(1)1s(2) \begin{cases} \alpha(1)\alpha(2) & - symmetric \\ \alpha(1)\beta(2) & - unsymmetric \\ \beta(1)\alpha(2) & - unsymmetric \\ \beta(1)\beta(2) & - symmetric \end{cases}$$

Similar to what we did for the unsymmetric space wavefunctions we need to make linear combinations in to form correct wavefunctions:

$$1s(1)1s(2) \begin{cases} \alpha(1)\alpha(2) & - symmetric \\ \dfrac{1}{\sqrt{2}}\left[\alpha(1)\beta(2)+\beta(1)\alpha(2)\right] & - symmetric \\ \dfrac{1}{\sqrt{2}}\left[\alpha(1)\beta(2)-\beta(1)\alpha(2)\right] & - antisymmetric \\ \beta(1)\beta(2) & - symmetric \end{cases}$$

Instead of one wavefunction we now have four different wavefunctions that all obey the correct exchange properties, where three of these are symmetric and only one that is antisymmetric. Experimentally we know that the ground state of He is a singlet. This indicates that the wavefunction should be antisymmetric under electron exchange. In general experimental evidence showed that for electrons only the antisymmetric wavefunction occur. This lead Pauli to postulate his now famous principle, here given in its most general form.

> The total wavefunction must by antisymmetric under the interchange of any pair of identical fermions and symmetrical under the interchanges of any two pair of identical bosons.

where fermions are spin half-integer particles (electrons, protons) and bosons are spin integer particles (photons, α -particles). Lets consider the consequences of the Pauli principle by considering the ground state of Li and for a minute forget what we already know. Similar to He we assume that wc can write the wavefunction as a product of hydrogen-like functions as $1s^3$,

$$\psi(Li) = \begin{cases} \alpha(1)\alpha(2)\alpha(3) & - symmetric \\ \alpha(1)\alpha(2)\beta(3) & - unsymmetric \\ \alpha(1)\beta(2)\alpha(3) & - unsymmetric \\ \beta(1)\alpha(2)\alpha(3) & - unsymmetric \\ \alpha(1)\beta(2)\beta(3) & - unsymmetric \\ \beta(1)\beta(2)\alpha(3) & - unsymmetric \\ \beta(2)\beta(2)\beta(3) & - unsymmetric \\ \beta(1)\beta(2)\beta(3) & - symmetric \end{cases}$$

Is is possible to find any linear combinations that are totally antisymmetric? No, which is in good agreement with our experimental observation that no states have $1s^3$ electronic configuration. If this was possible we would not have a periodic table! Therefore, from the Pauli exclusion principle we see that no two electrons and occupy the same spin-orbital. For Li the lowest function most therefore have a space wavefunction of the form 1s(1)1s(2)2s(3).

Slater-determinants

This procedure for generate wavefunctions that exhibit the correct symmetry easily becomes very tedious. Slater realized that writing the wavefunction as a determinant was an easy way of generating wavefunctions that are antisymmetric. A 2×2 determinant is given by:

$$\begin{vmatrix} a_{11} & a_{12} \\ a_{21} & a_{22} \end{vmatrix} = a_{11}a_{22} - a_{12}a_{21}$$

For a general A($n \times n$) determinant we need to use the cofactor (or Laplace expansion of the determinant as,

$$\det(A) = \sum_{k=1}^{n} (-1)^{i+k} a_{ik} \det(A_{ik}),$$

where A_{jk} is the determinant that remain after removing the ith row and the kth column of A These determinant are now known as Slater determinants. The properties of determinant ensures the Pauli principle since 1) change sign when interchange two rows (or columns), 2) if two rows are identical the determinant is zero, i.e. no two electrons can be in the same spin-orbital and 3) if a row is zero then the determinant is zero. The general form for a N electron wavefuntion is,

$$\psi_a^N = \frac{1}{\sqrt{N!}} \begin{vmatrix} \phi_1(x_1) & \phi_2(x_1) & \dots & \phi_N(x_1) \\ \phi_1(x_2) & \phi_2(x_2) & \dots & \phi_N(x_2) \\ \vdots & \vdots & \ddots & \vdots \\ \phi_1(x_N) & \phi_2(x_N) & \dots & \phi_N(x_N) \end{vmatrix},$$

where ϕ_i is a general spin-orbital. Therefore, for Li the slater determinant would be,

$$\psi(Li) = \frac{1}{\sqrt{6}} \begin{vmatrix} 1s(1)\alpha(1) & 1s(1)\beta(1) & 2s(1)\alpha(1) \\ 1s(1)\alpha(1) & 1s(2)\beta(2) & 2s(2)\alpha(2) \\ 1s(3)\alpha(3) & 1s(3)\beta(3) & 2s(3)\alpha(3) \end{vmatrix}$$

$$= \frac{1}{\sqrt{6}} \begin{vmatrix} 1s(1) & \overline{1s}(1) & 2s(1) \\ 1s(2) & \overline{1s}(2) & 2s(2) \\ 1s(3) & \overline{1s}(3) & 2s(3) \end{vmatrix}.$$

Since we can always construct the full Slater determinant if we know a list of all the spin-orbitals a simplified notation is:

$$\psi = \left| \phi_1 \phi_2 \phi_3 .... \phi_n \right\rangle.$$

Example: Write the normalized Slater determinant for the beryllium in the $1s^2 2s^2$ configurations. Explain how you would expand it.

$$\psi(Be) = \frac{1}{\sqrt{4!}} \begin{vmatrix} 1s(1) & \overline{1s}(1) & 2s(1) & \overline{2s}(1) \\ 1s(2) & \overline{1s}(2) & 2s(2) & \overline{2s}(2) \\ 1s(3) & \overline{1s}(3) & 2s(3) & \overline{2s}(3) \\ 1s(4) & \overline{1s}(4) & 2s(4) & \overline{2s}(4) \end{vmatrix}$$

The Hamiltonian and Spin

Since the hamiltonian is independent of spin, at the non-relativitis level of theory, we know that it commutes with the spin operators:

$$\left[\hat{H}, \hat{S}_z \right] = \left[\hat{H}, \hat{S}^2 \right] = \left[\hat{S}^2, \hat{S}_z \right] = 0$$

thus, we can specify simultaneous eigenfunctions of these operators. What about the energies:

$$\hat{H} \Psi(x,y,z,\sigma) = \left[\hat{H} \psi(x,y,z) \right] g(\sigma) = \left[E \psi(x,y,z) \right] g(\sigma) = E \Psi(x,y,z,\sigma)$$

and our energies are the same as we previously found.

Singlet and Triplet Excited States of Helium

For the 1s2s state of He we saw that there are two different space functions that have the correct symmetry, one symmetric and one symmetric. We have also learned that for an electron it is the total wavefunction that needs to be antisymmetric with respect to the interchange of both space and spin labels. Therefore, we need to combine the symmetric space part with an antisymmetric spin part and a symmetric spin part with the antisymmetric space functions. This gives us for the 1s2s:

$$\psi_{s,a}(He) = \frac{1}{\sqrt{2}} \left[1s2s + 2s1s \right] \frac{1}{\sqrt{2}} \left[\alpha\beta - \beta\alpha \right],$$

called a singlet state, and,

$$\psi_{s,a}(He) = \frac{1}{\sqrt{2}} \left[1s2s + 2s1s \right] \begin{cases} \alpha\alpha \\ \frac{1}{\sqrt{2}} \left[\alpha\beta + \beta\alpha \right], \\ \beta\beta \end{cases}$$

called a triplet state. Since all four state satisfy the correct symmetry and are linear independent it indicates that there are 4 distinct physical states. Is it possible to write a single Slater determinant for each of these states? The answer is no and is very important to realize. It is not always possible with only one Slater determinant to write a wave function that has the correct symmetry of the full wavefunctions,

$$\psi_{s,a}(He) = \frac{1}{\sqrt{2}}\left\{ \frac{1}{\sqrt{2}}\begin{vmatrix} 1s & \overline{2s} \\ 1s & 2s \end{vmatrix} - \frac{1}{\sqrt{2}}\begin{vmatrix} \overline{1s} & 2s \\ 1s & 2s \end{vmatrix} \right\},$$

but instead we need to make linear combinations of Slater determinants. All four state are degenerate with the independent electron model, would we expect this to be the case for the full Hamiltonian?

Let evaluate the energy of these state using the full Hamiltonian. What is the expression for the full Hamiltonian? and how do we evaluate the energy?

$$\langle E \rangle = \int \psi^* \widehat{H}\psi$$

$$= \int \psi^* \left[\hat{h}_1 + \hat{h}_2 + \frac{1}{r_{12}} \right] \psi \, dr$$

$$= \int \psi^* \left[-\frac{1}{2}\nabla_1^2 - \frac{1}{2}\nabla_2^2 - \frac{2}{R_1} - \frac{2}{R_2} + \frac{1}{r_{12}} \right] \psi \, dr$$

Since the Hamiltonian do not dependent on spin we can integrate the spin out first. This means that the energy is completely determined by the space functions. We can now substitute the space function for the triplet and singlet case in the energy expression,

$$\langle E1,3 \rangle = \int\int [1s*2s \pm 2s*1s*] \left[\hat{h}_1 + \hat{h}_2 + \frac{1}{r_{12}} \right] [1s2s \pm 2s1s] \, d1d2,$$

using that the orbitals are ortho-normal we can simplify the integrals,

$$\langle E1,3 \rangle = \int 1s^* \hat{h}_1 1s d1 + \int 2s^* \hat{h}_1 2s d1$$

$$+ \int\int 1s^* 2s^* \frac{1}{r^{12}} 1s2s d1d2$$

$$\pm \int\int 1s^* 2s^* \frac{1}{r_{12}} 2s1s d1d2$$

$$= E_{1s} + E_{2s} + J \pm K.$$

where J is the Coulomb integral and K is the exchange integral. J represent the Coulomb repulsion between two electrons where K is similar but with one indices exchanged. Therefore, the triples state is lower in energy than the singlet since K (and J) is positive. This is in agreement with our experimental observation of the the 1s2s configurations for which two states are found and the lowest split into 3 state in a magnetic field.

Angular Momentum in Many-electron Atoms

When dealing with many-electron systems we need to account for the spin and orbital angular momentum of the electrons. We therefore need to understand the rules for adding angular momentum. For many electron atom the individual angular momentum operators do not compute with the Hamiltonian, however, their sum does. The total orbital angular momentum is defined as the vector sum of the orbital angular momenta of the individual electrons:

$$L = \sum_i L_i,$$

with eigenvalues,

$$\hat{L}^2 \left| LM_L \right\rangle = \hbar L(L+1) \left| LM_L \right\rangle,$$

and for the projection,

$$\hat{L}_z \left| LM_L \right\rangle = \hbar M_L \left| LM_L \right\rangle,$$

n general the addition of two angular momenta with quantum number l_1 and l_2 will results in a total quantum number whose number J has the following possible values,

$$L = \left| l_1 + l_2 \right|, \left| l_1 + l_2 \right|, \dots \left| l_1 - l_2 \right|,$$

And,

$$M = m_1 + m_2.$$

The total orbital angular momentum quantum number L of an atom is denoted by S for L = 0, P for L = 1, D for L = 2, F for L = 3, and so forth.

Ladder Operators

Similar for S we have,

$$\hat{S}^2 \left| S M_s \right\rangle = \hbar^2 S(S+1) \left| S M_s \right\rangle,$$

and for the projection,

$$\hat{S}_z \left| S M_s \right\rangle = \hbar M_s \left| S M_s \right\rangle.$$

So far we have seen how to work with $\hat{S}^2$ and $\hat{S}_z$ operators. Similar to what we did for the orbital angular momentum we can define ladder operators as,

$$\hat{S}_+ = \hat{S}_x + i\hat{S}_y,$$

for the raising operator and,

$$\hat{S}_- = \hat{S}_x - i\hat{S}_y,$$

for the lowering operator. Operating with these operators on our spin functions gives,

$$\hat{S}_+\beta = \hbar\alpha$$
$$\hat{S}_+\alpha = 0$$
$$\hat{S}_-\beta = 0$$
$$\hat{S}_-\alpha = \hbar\beta.$$

Using these we can now operate with the other two components on the functions and get,

$$\hat{S}_x\beta = 1/2\left(\hat{S}_+ + \hat{S}_-\right)\beta = 1/2\hbar\alpha$$
$$\hat{S}_y\beta = 1(i2)\left(\hat{S}_+ - \hat{S}_-\right)\beta = -i/2\hbar\alpha$$
$$\hat{S}_x\alpha = 1/2\hbar\beta$$
$$\hat{S}_y\alpha = i/2\hbar\beta.$$

Eigenvalues of a two-electron Spin Function

The excited state configuration of He (1s2s) have the following possible spin eigenfunctions:

$$\alpha(1)\alpha(2)$$
$$\frac{1}{\sqrt{2}}\left[\alpha(1)\beta(2) + \beta(1)\alpha(2)\right]$$
$$\beta(1)\beta(2)\frac{1}{\sqrt{2}}\left[\alpha(1)\beta(2) - \beta(1)\alpha(2)\right].$$

Where the possible total spin is $S = (1/2 + 1/2),(1/2 - 1/2 = (1, 0)$ with degeneracy $d = (2L + 1)(2S + 1)$ which gives 3 and 1, respectively. Thus, the term symbols are ${}^3S_1$ and ${}^1S_0$, respectively. Now lets use the spin operators and verify this. First lets start with the projection:

$$\hat{S}_z\alpha(1)\alpha(2) = \hat{S}_{1z}\alpha(1)\alpha(2) + \hat{S}_{2z}\alpha(1)\alpha(2)$$
$$= \frac{1}{2}\hbar\alpha(1)\alpha(2) + \frac{1}{2}\hbar\alpha(1)\alpha(2)$$
$$= \hbar\alpha(1)\alpha(2).$$

Similar for the other functions we find:

$$\hat{S}_z\beta(1)\beta(2) = -\hbar\beta(1)\beta(2)$$
$$\hat{S}_z\left[\alpha(1)\beta(2) \pm \beta(1)\alpha(2)\right] = 0.$$

For $\hat{S}^2$ we get:

$$\hat{S}^2 = \left(\hat{S}_1 + \hat{S}_1\right)\cdot\left(\hat{S}_1 + \hat{S}_1\right) = \hat{S}_1^2 + \hat{S}_2^2 + 2\left(\hat{S}_{1x}\hat{S}_{2x} + \hat{S}_{1y}\hat{S}_{2y} + \hat{S}_{1z}\hat{S}_{2z}\right).$$

Operating with this on the eigenfunctions give:

$$\hat{S}^2\alpha(1)\alpha(2) = \alpha(2)\hat{S}_1^2\alpha(1) + \alpha(1)\hat{S}_2^2\alpha(2) + 2\hat{S}_{1x}\alpha(1)\hat{S}_{2x}\alpha(2)$$
$$+ \hat{S}_{1y}\hat{S}_{2y} + \hat{S}_{1z}\hat{S}_{2z}$$
$$= \left(3/4\hbar^2 + 3/4\hbar^2 + 1/2\hbar^2\right)\alpha(1)\alpha(2) = 2\hbar^2\alpha(2).$$

Similarly for the other eigenfunctions we get:

$$\hat{S}^2\beta(1)\beta(2) = 2\hbar^2\beta(1)\beta(2),$$
$$\hat{S}^2\left[\alpha(1)\beta(2) + \beta(1)\alpha(2)\right] = 2\hbar\left[\alpha(1)\beta(2) + \beta(1)\alpha(2)\right]\hat{S}^2\left[\alpha(1)\beta(2) - \beta(1)\alpha(2)\right] = 0.$$

Therefore, we see that for the singlet we indeed get S = 0, ms = 0 and for

all of the triplet states we S = 1, m_s = −1, 0, 1 as expected.

Term Symbols

Example: Find the possible quantum number L for states of the Carbon atom with the following electron configuration $1s^2 2s^2 2p3p$? Since the s electrons have zero angular momentum they do not contribute. The 2p has $l = 1$ and the 3p $l = 1$ gives L from $1+1 = 2$ to $|1-1| = 0$, therefore L = 0, 1, 2.

Example: Find the possible values for S for the states arising from $1s^2 2s^2 2p3p$. S electrons will contribute nothing due to Pauli principle, i.e. $1/2 - 1/2 = 0$. For the last two electrons we can get S = 0 and S = 1.

For a given S we have $2S + 1$ values for MS where $2S + 1$ is called the multiplicity. For $2S + 1 = 1, 2, 3, 4$ is called singlet, doublet, triplet, quartet. The total angular momentum is then:

$$J = L + S$$

The allow us to characterize the different electronic states in a multi-electron atoms using what is know as a term symbol,

Where,

- S is the total spin quantum number. $2S + 1$ is the spin multiplicity: the maximum number of different possible states of J for a given (L, S) combination.

- L is the total orbital quantum number in spectroscopic notation. The symbols for L = 0, 1, 2, 3, 4, 5 are S, P, D, F, G, H respectively.

- J is the total angular momentum quantum number.

for which a total number of states are given by $(2S + 1)(2L + 1)$. An alternative statement of Hund's rule is that the term with the highest multiplicity is lowest in energy. Remember that Hund's rule works well for the ground state, however, can fail for excited states.

Example: Write down the possible term symbols for Carbon atom in the $1s^2 2s^2 2p3p$ configuration.

$$L = 2, \ S = 1 \ gives \ {}^3D_3, \ {}^3D_2, \ {}^3D_1,$$
$$L = 2, \ S = 0 \ gives \ {}^1D_2,$$
$$L = 1, \ S = 1 \ gives \ {}^3P_2, \ {}^3P_1, \ {}^3P_0,$$
$$L = 1, \ S = 0 \ gives \ {}^1P_1,$$
$$L = 0, \ S = 1 \ gives \ {}^3S_1, \ and$$
$$L = 0, \ S = 0 \ gives \ {}^1S_0.$$

The Atomic Hamiltonian

Although our non-relativistic hamiltonian do not include spin, there is a small term in the true Hamiltonian which comes from the interactions between the spin and orbital angulr momentum and is referred to as spin-orbit interactions. The atomic hamiltonian is then given by a sum of three terms:

$$\widehat{H} = \widehat{H}_0 + \widehat{H}_{rep} + \widehat{H}_{S.O},$$

where the first term is a sum of hydrogen-like Hamiltonians,

$$\widehat{H}_0 = \sum_{i=1}^{n} \left(-\frac{1}{2}\nabla^2 - \frac{Z}{r_i} \right),$$

the second term is the electron-electron repulsion term,

$$\widehat{H}_{rep} = \frac{1}{2}\sum_{i,j}^{n}{}' \frac{1}{r_{i,j}},$$

and the third term is the spin-orbit interactions,

$$\widehat{H}_{S.O} = \sum_{i=1}^{n} \xi \widehat{L}_i \cdot \widehat{S}_i.$$

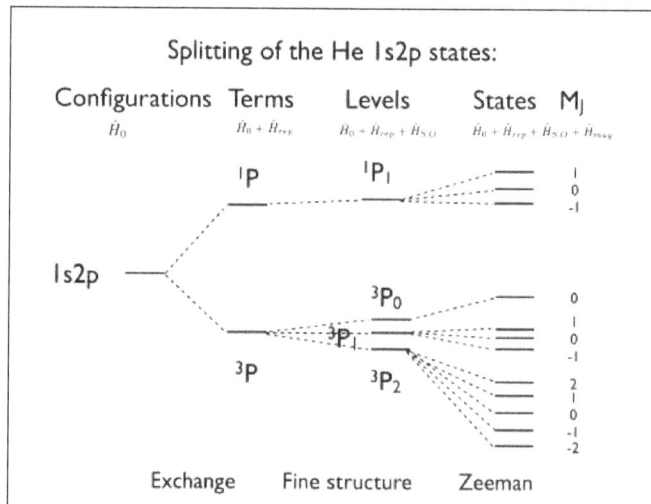

Splitting of the He 1s2p states:

The repulsion operator splits antisymmetric and symmetric space wavefunctions into terms and the spin-orbitat interactions splits the terms into the individual levels, resulting in the fine-structure of electronic spectra. Finally we can use a magnetic field perturbation to split the individual levels in states. The magnetic field perturbation is given

$$\widehat{H}_{mag} = \beta_e \left(\widehat{J} + \widehat{S} \right) \cdot B = \beta_e \left(\widehat{J}_z + \widehat{S}_z \right)$$

and is known as the Zeeman effect.

References

- Atom, science: britannica.com, retrieved 15 June, 2019

- "the rutherford experiment". Rutgers university. Archived from the original on november 14, 2001. Retrieved february 26, 2013

- Atomic-mass, supplemental-modules-(physical-and-theoretical-chemistry), physical-and-theoretical-chemistry-textbook-maps,bookshelves: libretexts.org, retrieved 15 January, 2019

- Palmer, d. (13 september 1997). "hydrogen in the universe". Nasa. Archived from the original on 29 october 2014. Retrieved 23 february 2017

- What-is-atomic-number: sciencing.com, retrieved 21 April, 2019

- Meija, juris; et al. (2016). "atomic weights of the elements 2013 (iupac technical report)". Pure and applied chemistry. 88 (3): 265–91. Doi:10.1515/pac-2015-0305

Fundamentals of Atomic Physics

The scientific study of atoms and their structures, arrangements, their energy states and interactions with other particles is termed as atomic physics. Some of its components include atomic units, atomic models, isotopes, isobars, atomic spectroscopy etc. All the diverse components of atomic physics have been carefully analysed in this chapter.

Atomic physics is the scientific study of the structure of the atom, its energy states, and its interactions with other particles and with electric and magnetic fields. Atomic physics has proved to be a spectacularly successful application of quantum mechanics, which is one of the cornerstones of modern physics.

The notion that matter is made of fundamental building blocks dates to the ancient Greeks, who speculated that earth, air, fire, and water might form the basic elements from which the physical world is constructed. They also developed various schools of thought about the ultimate nature of matter. Perhaps the most remarkable was the atomist school founded by the ancient Greeks Leucippus of Miletus and Democritus of Thrace about 440 BC. For purely philosophical reasons, and without benefit of experimental evidence, they developed the notion that matter consists of indivisible and indestructible atoms. The atoms are in ceaseless motion through the surrounding void and collide with one another like billiard balls, much like the modern kinetic theory of gases. However, the necessity for a void (or vacuum) between the atoms raised new questions that could not be easily answered. For this reason, the atomist picture was rejected by Aristotle and the Athenian school in favour of the notion that matter is continuous. The idea nevertheless persisted, and it reappeared 400 years later in the writings of the Roman poet Lucretius, in his work De rerum natura (On the Nature of Things).

Little more was done to advance the idea that matter might be made of tiny particles until the 17th century. The English physicist Isaac Newton, in his Principia Mathematica (1687), proposed that Boyle's law, which states that the product of the pressure and the volume of a gas is constant at the same temperature, could be explained if one assumes that the gas is composed of particles. In 1808 the English chemist John Dalton suggested that each element consists of identical atoms, and in 1811 the Italian physicist Amedeo Avogadro hypothesized that the particles of elements may consist of two or more atoms stuck together. Avogadro called such conglomerations molecules, and, on the basis of experimental work, he conjectured that the molecules in a gas of hydrogen or oxygen are formed from pairs of atoms.

During the 19th century there developed the idea of a limited number of elements, each consisting of a particular type of atom, that could combine in an almost limitless number of ways to form chemical compounds. At mid-century the kinetic theory of gases successfully attributed such phenomena as the pressure and viscosity of a gas to the motions of atomic and molecular particles. By 1895 the growing weight of chemical evidence and the success of the kinetic theory left little doubt that atoms and molecules were real.

The internal structure of the atom, however, became clear only in the early 20th century with the work of the British physicist Ernest Rutherford and his students. Until Rutherford's efforts, a popular model of the atom had been the so-called "plum-pudding" model, advocated by the English physicist Joseph John Thomson, which held that each atom consists of a number of electrons (plums) embedded in a gel of positive charge (pudding); the total negative charge of the electrons exactly balances the total positive charge, yielding an atom that is electrically neutral. Rutherford conducted a series of scattering experiments that challenged Thomson's model. Rutherford observed that when a beam of alpha particles (which are now known to be helium nuclei) struck a thin gold foil, some of the particles were deflected backward. Such large deflections were inconsistent with the plum-pudding model.

This work led to Rutherford's atomic model, in which a heavy nucleus of positive charge is surrounded by a cloud of light electrons. The nucleus is composed of positively charged protons and electrically neutral neutrons, each of which is approximately 1,836 times as massive as the electron. Because atoms are so minute, their properties must be inferred by indirect experimental techniques. Chief among these is spectroscopy, which is used to measure and interpret the electromagnetic radiation emitted or absorbed by atoms as they undergo transitions from one energy state to another. Each chemical element radiates energy at distinctive wavelengths, which reflect their atomic structure. Through the procedures of wave mechanics, the energies of atoms in various energy states and the characteristic wavelengths they emit may be computed from certain fundamental physical constants—namely, electron mass and charge, the speed of light, and Planck's constant. Based on these fundamental constants, the numerical predictions of quantum mechanics can account for most of the observed properties of different atoms. In particular, quantum mechanics offers a deep understanding of the arrangement of elements in the periodic table, showing, for example, that elements in the same column of the table should have similar properties.

In recent years the power and precision of lasers have revolutionized the field of atomic physics. On the one hand, lasers have dramatically increased the precision with which the characteristic wavelengths of atoms can be measured. For example, modern standards of time and frequency are based on measurements of transition frequencies in atomic cesium and the definition of the metre as a unit of length is now related to frequency measurements through the velocity of light. In addition, lasers have made possible entirely new technologies for isolating individual atoms in electromagnetic traps and cooling them to near absolute zero. When the atoms are brought essentially to rest in the trap, they can undergo a quantum mechanical phase transition to form a superfluid known as a Bose-Einstein condensation, while remaining in the form of a dilute gas. In this new state of matter, all the atoms are in the same coherent quantum state. As a consequence, the atoms lose their individual identities, and their quantum mechanical wavelike properties become dominant. The entire condensate then responds to external influences as a single coherent entity (like a school of fish), instead of as a collection of individual atoms. Recent work has shown that a coherent beam of atoms can be extracted from the trap to form an "atom laser" analogous to the coherent beam of photons in a conventional laser. The atom laser is still in an early stage of development, but it has the potential to become a key element of future technologies for the fabrication of microelectronic and other nanoscale devices.

SUB-ATOMIC PARTICLES

Electron

The electron is a subatomic particle, symbol e^- or β^-, whose electric charge is negative one elementary charge. Electrons belong to the first generation of the lepton particle family, and are generally thought to be elementary particles because they have no known components or substructure. The electron has a mass that is approximately 1/1836 that of the proton. Quantum mechanical properties of the electron include an intrinsic angular momentum (spin) of a half-integer value, expressed in units of the reduced Planck constant, $\hbar$. Being fermions, no two electrons can occupy the same quantum state, in accordance with the Pauli exclusion principle. Like all elementary particles, electrons exhibit properties of both particles and waves: they can collide with other particles and can be diffracted like light. The wave properties of electrons are easier to observe with experiments than those of other particles like neutrons and protons because electrons have a lower mass and hence a longer de Broglie wavelength for a given energy.

Electrons play an essential role in numerous physical phenomena, such as electricity, magnetism, chemistry and thermal conductivity, and they also participate in gravitational, electromagnetic and weak interactions. Since an electron has charge, it has a surrounding electric field, and if that electron is moving relative to an observer, said observer will observe it to generate a magnetic field. Electromagnetic fields produced from other sources will affect the motion of an electron according to the Lorentz force law. Electrons radiate or absorb energy in the form of photons when they are accelerated. Laboratory instruments are capable of trapping individual electrons as well as electron plasma by the use of electromagnetic fields. Special telescopes can detect electron plasma in outer space. Electrons are involved in many applications such as electronics, welding, cathode ray tubes, electron microscopes, radiation therapy, lasers, gaseous ionization detectors and particle accelerators.

Interactions involving electrons with other subatomic particles are of interest in fields such as chemistry and nuclear physics. The Coulomb force interaction between the positive protons within atomic nuclei and the negative electrons without, allows the composition of the two known as atoms. Ionization or differences in the proportions of negative electrons versus positive nuclei changes the binding energy of an atomic system. The exchange or sharing of the electrons between two or more atoms is the main cause of chemical bonding. In 1838, British natural philosopher Richard Laming first hypothesized the concept of an indivisible quantity of electric charge to explain the chemical properties of atoms. Irish physicist George Johnstone Stoney named this charge 'electron' in 1891, and J. J. Thomson and his team of British physicists identified it as a particle in 1897. Electrons can also participate in nuclear reactions, such as nucleosynthesis in stars, where they are known as beta particles. Electrons can be created through beta decay of radioactive isotopes and in high-energy collisions, for instance when cosmic rays enter the atmosphere. The antiparticle of the electron is called the positron; it is identical to the electron except that it carries electrical and other charges of the opposite sign. When an electron collides with a positron, both particles can be annihilated, producing gamma ray photons.

Discovery of Effect of Electric Force

The ancient Greeks noticed that amber attracted small objects when rubbed with fur. Along with lightning, this phenomenon is one of humanity's earliest recorded experiences with electricity. In his 1600 treatise De Magnete, the English scientist William Gilbert coined the New Latin term electrica, to refer to those substances with property similar to that of amber which attract small objects after being rubbed.

Discovery of Two Kinds of Charges

In the early 1700s, French chemist Charles François du Fay found that if a charged gold-leaf is repulsed by glass rubbed with silk, then the same charged gold-leaf is attracted by amber rubbed with wool. From this and other results of similar types of experiments, du Fay concluded that electricity consists of two electrical fluids, vitreous fluid from glass rubbed with silk and resinous fluid from amber rubbed with wool. These two fluids can neutralize each other when combined. American scientist Ebenezer Kinnersley later also independently reached the same conclusion. A decade later Benjamin Franklin proposed that electricity was not from different types of electrical fluid, but a single electrical fluid showing an excess (+) or deficit (-). He gave them the modern charge nomenclature of positive and negative respectively. Franklin thought of the charge carrier as being positive, but he did not correctly identify which situation was a surplus of the charge carrier, and which situation was a deficit.

Between 1838 and 1851, British natural philosopher Richard Laming developed the idea that an atom is composed of a core of matter surrounded by subatomic particles that had unit electric charges. Beginning in 1846, German physicist William Weber theorized that electricity was composed of positively and negatively charged fluids, and their interaction was governed by the inverse square law. After studying the phenomenon of electrolysis in 1874, Irish physicist George Johnstone Stoney suggested that there existed a "single definite quantity of electricity", the charge of a monovalent ion. He was able to estimate the value of this elementary charge e by means of Faraday's laws of electrolysis. However, Stoney believed these charges were permanently attached to atoms and could not be removed. In 1881, German physicist Hermann von Helmholtz argued that both positive and negative charges were divided into elementary parts, each of which "behaves like atoms of electricity".

Stoney initially coined the term electrolion in 1881. Ten years later, he switched to electron to describe these elementary charges, writing in 1894: "an estimate was made of the actual amount of this most remarkable fundamental unit of electricity, for which I have since ventured to suggest the name electron". A 1906 proposal to change to electrion failed because Hendrik Lorentz preferred to keep electron. The word electron is a combination of the words electric and ion. The suffix -on which is now used to designate other subatomic particles, such as a proton or neutron, is in turn derived from electron.

Discovery of Free Electrons Outside Matter

The discovery of electron by Joseph Thomson was closely tied with the experimental and theoretical research of cathode rays for decades by many physicists. While studying electrical conductivity in rarefied gases in 1859, The German physicist Julius Plucker observed that the phosphorescent light, which was caused by radiation emitted from the cathode, appeared at the tube wall near the cathode, and the region of the phosphorescent light could be moved by application of a magnetic field. In 1869, Plucker's student Johann Wilhelm Hittorf found that a solid body placed in between the cathode and the phosphorescence would cast a shadow upon the phosphorescent region of the tube. Hittorf inferred that there are straight rays emitted from the cathode and that the phosphorescence was caused by the rays striking the tube walls. In 1876, the German physicist Eugen Goldstein showed that the rays were emitted perpendicular to the cathode surface, which distinguished between the rays that were emitted from the cathode and the incandescent light. Goldstein dubbed the rays cathode rays.

During the 1870s, the English chemist and physicist Sir William Crookes developed the first cathode ray tube to have a high vacuum inside. He then showed in 1874 that the cathode rays can turn a small paddle wheel when placed in their path. Therefore, he concluded that the rays carried momentum. Furthermore, by applying a magnetic field, he was able to deflect the rays, thereby demonstrating that the beam behaved as though it were negatively charged. In 1879, he proposed that these properties could be explained by regarding cathode rays as composed of negatively charged gaseous molecules in fourth state of matter in which the mean free path of the particles is so long that collisions may be ignored.

The German-born British physicist Arthur Schuster expanded upon Crookes' experiments by placing metal plates parallel to the cathode rays and applying an electric potential between the plates. The field deflected the rays toward the positively charged plate, providing further evidence that the rays carried negative charge. By measuring the amount of deflection for a given level of current, in 1890 Schuster was able to estimate the charge-to-mass ratio of the ray components. However, this produced a value that was more than a thousand times greater than what was expected, so little credence was given to his calculations at the time.

In 1892 Hendrik Lorentz suggested that the mass of these particles (electrons) could be a consequence of their electric charge.

While studying naturally fluorescing minerals in 1896, the French physicist Henri Becquerel discovered that they emitted radiation without any exposure to an external energy source. These radioactive materials became the subject of much interest by scientists, including the New Zealand physicist Ernest Rutherford who discovered they emitted particles. He designated these particles alpha and beta, on the basis of their ability to penetrate matter. In 1900, Becquerel showed that the beta rays emitted by radium could be deflected by an electric field, and that their mass-to-charge ratio was the same as for cathode rays. This evidence strengthened the view that electrons existed as components of atoms.

In 1897, the British physicist J. J. Thomson, with his colleagues John S. Townsend and H. A. Wilson, performed experiments indicating that cathode rays really were unique particles, rather than waves, atoms or molecules as was believed earlier. Thomson made good estimates of both the charge e and the mass m, finding that cathode ray particles, which he called "corpuscles," had perhaps one thousandth of the mass of the least massive ion known: hydrogen. He showed that their charge-to-mass ratio, e/m, was independent of cathode material. He further showed that the negatively charged particles produced by radioactive materials, by heated materials and by illuminated materials were universal. The name electron was adoted for these particles by the scientific community, mainly due to the advocation by G. F. Fitzgerald, J. Larmor, and H. A. Lorenz.

The electron's charge was more carefully measured by the American physicists Robert Millikan and Harvey Fletcher in their oil-drop experiment of 1909, the results of which were published in 1911. This experiment used an electric field to prevent a charged droplet of oil from falling as a result of gravity. This device could measure the electric charge from as few as 1–150 ions with an error margin of less than 0.3%. Comparable experiments had been done earlier by Thomson's team, using clouds of charged water droplets generated by electrolysis, and in 1911 by Abram Ioffe, who independently obtained the same result as Millikan using charged microparticles of metals, then published his results in 1913. However, oil drops were more stable than water drops because of their slower evaporation rate, and thus more suited to precise experimentation over longer periods of time.

Around the beginning of the twentieth century, it was found that under certain conditions a fast-moving charged particle caused a condensation of supersaturated water vapor along its path. In 1911, Charles Wilson used this principle to devise his cloud chamber so he could photograph the tracks of charged particles, such as fast-moving electrons.

A beam of electrons deflected in a circle by a magnetic field.

J. J. Thomson.   Robert Millikan.

Atomic Theory

By 1914, experiments by physicists Ernest Rutherford, Henry Moseley, James Franck and Gustav Hertz had largely established the structure of an atom as a dense nucleus of positive charge surrounded by lower-mass electrons. In 1913, Danish physicist Niels Bohr postulated that electrons resided in quantized energy states, with their energies determined by the angular momentum of the electron's orbit about the nucleus. The electrons could move between those states, or orbits, by the emission or absorption of photons of specific frequencies. By means of these quantized orbits, he accurately explained the spectral lines of the hydrogen atom. However, Bohr's model failed to account for the relative intensities of the spectral lines and it was unsuccessful in explaining the spectra of more complex atoms.

Chemical bonds between atoms were explained by Gilbert Newton Lewis, who in 1916 proposed that a covalent bond between two atoms is maintained by a pair of electrons shared between them. Later, in 1927, Walter Heitler and Fritz London gave the full explanation of the electron-pair formation and chemical bonding in terms of quantum mechanics. In 1919, the American chemist Irving Langmuir elaborated on the Lewis' static model of the atom and suggested that all electrons were distributed in successive "concentric (nearly) spherical shells, all of equal thickness". In turn, he divided the shells into a number of cells each of which contained one pair of electrons. With this model Langmuir was able to qualitatively explain the chemical properties of all elements in the periodic table, which were known to largely repeat themselves according to the periodic law.

In 1924, Austrian physicist Wolfgang Pauli observed that the shell-like structure of the atom could be explained by a set of four parameters that defined every quantum energy state, as long as each state was occupied by no more than a single electron. This prohibition against more than one electron occupying the same quantum energy state became known as the Pauli exclusion principle. The physical mechanism to explain the fourth parameter, which had two distinct possible values, was provided by the Dutch physicists Samuel Goudsmit and George Uhlenbeck. In 1925, they suggested that an electron, in addition to the angular momentum of its orbit, possesses an intrinsic angular momentum and magnetic dipole moment. This is analogous to the rotation of the Earth on its axis as it orbits the Sun. The intrinsic angular momentum became known as spin, and explained the previously mysterious splitting of spectral lines observed with a high-resolution spectrograph; this phenomenon is known as fine structure splitting.

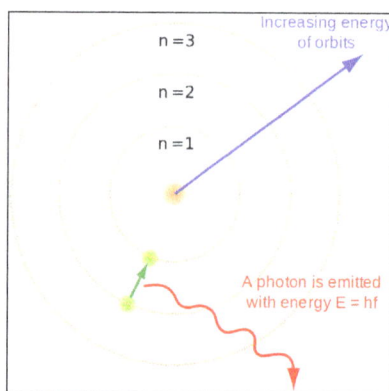

The Bohr model of the atom, showing states of electron with energy quantized by the number n.
An electron dropping to a lower orbit emits a photon equal to the energy difference between the orbits.

Quantum Mechanics

In his 1924 dissertation Recherches sur la théorie des quanta (Research on Quantum Theory), French physicist Louis de Broglie hypothesized that all matter can be represented as a de Broglie wave in the manner of light. That is, under the appropriate conditions, electrons and other matter would show properties of either particles or waves. The corpuscular properties of a particle are demonstrated when it is shown to have a localized position in space along its trajectory at any given moment. The wave-like nature of light is displayed, for example, when a beam of light is passed through parallel slits thereby creating interference patterns. In 1927 George Paget Thomson, discovered the interference effect was produced when a beam of electrons was passed through thin metal foils and by American physicists Clinton Davisson and Lester Germer by the reflection of electrons from a crystal of nickel.

De Broglie's prediction of a wave nature for electrons led Erwin Schrödinger to postulate a wave equation for electrons moving under the influence of the nucleus in the atom. In 1926, this equation, the Schrödinger equation, successfully described how electron waves propagated. Rather than yielding a solution that determined the location of an electron over time, this wave equation also could be used to predict the probability of finding an electron near a position, especially a position near where the electron was bound in space, for which the electron wave equations did not change in time. This approach led to a second formulation of quantum mechanics (the first by Heisenberg in 1925), and solutions of Schrödinger's equation, like Heisenberg's, provided derivations of the energy states of an electron in a hydrogen atom that were equivalent to those that had been derived first by Bohr in 1913, and that were known to reproduce the hydrogen spectrum. Once spin and the interaction between multiple electrons were describable, quantum mechanics made it possible to predict the configuration of electrons in atoms with atomic numbers greater than hydrogen.

In 1928, building on Wolfgang Pauli's work, Paul Dirac produced a model of the electron – the Dirac equation, consistent with relativity theory, by applying relativistic and symmetry considerations to the hamiltonian formulation of the quantum mechanics of the electro-magnetic field. In order to resolve some problems within his relativistic equation, Dirac developed in 1930 a model of the vacuum as an infinite sea of particles with negative energy, later dubbed the Dirac sea. This led him to predict the existence of a positron, the antimatter counterpart of the electron. This particle was discovered in 1932 by Carl Anderson, who proposed calling standard electrons negatons and using electron as a generic term to describe both the positively and negatively charged variants.

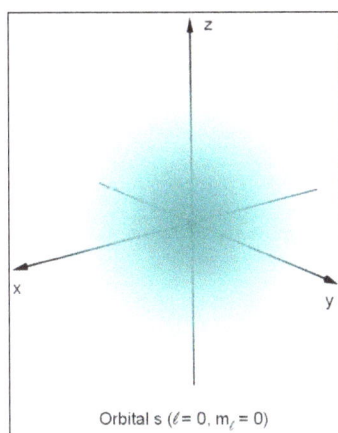

Orbital s ($\ell = 0$, $m_\ell = 0$)

In quantum mechanics, the behavior of an electron in an atom is described by an orbital, which is a probability distribution rather than an orbit. In the figure, the shading indicates the relative probability to "find" the electron, having the energy corresponding to the given quantum numbers, at that point.

In 1947 Willis Lamb, working in collaboration with graduate student Robert Retherford, found that certain quantum states of the hydrogen atom, which should have the same energy, were shifted in relation to each other; the difference came to be called the Lamb shift. About the same time,

Polykarp Kusch, working with Henry M. Foley, discovered the magnetic moment of the electron is slightly larger than predicted by Dirac's theory. This small difference was later called anomalous magnetic dipole moment of the electron. This difference was later explained by the theory of quantum electrodynamics, developed by Sin-Itiro Tomonaga, Julian Schwinger and Richard Feynman in the late 1940s.

Particle Accelerators

With the development of the particle accelerator during the first half of the twentieth century, physicists began to delve deeper into the properties of subatomic particles. The first successful attempt to accelerate electrons using electromagnetic induction was made in 1942 by Donald Kerst. His initial betatron reached energies of 2.3 MeV, while subsequent betatrons achieved 300 MeV. In 1947, synchrotron radiation was discovered with a 70 MeV electron synchrotron at General Electric. This radiation was caused by the acceleration of electrons through a magnetic field as they moved near the speed of light.

With a beam energy of 1.5 GeV, the first high-energy particle collider was ADONE, which began operations in 1968. This device accelerated electrons and positrons in opposite directions, effectively doubling the energy of their collision when compared to striking a static target with an electron. The Large Electron–Positron Collider (LEP) at CERN, which was operational from 1989 to 2000, achieved collision energies of 209 GeV and made important measurements for the Standard Model of particle physics.

Confinement of Individual Electrons

Individual electrons can now be easily confined in ultra small ($L = 20$ nm, $W = 20$ nm) CMOS transistors operated at cryogenic temperature over a range of −269 °C (4 K) to about −258 °C (15 K). The electron wavefunction spreads in a semiconductor lattice and negligibly interacts with the valence band electrons, so it can be treated in the single particle formalism, by replacing its mass with the effective mass tensor.

Characteristics

Classification

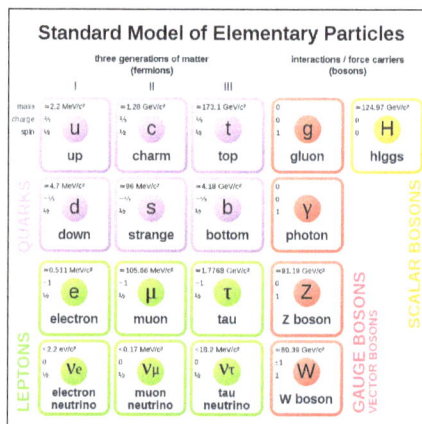

Standard Model of elementary particles. The electron (symbol e) is on the left.

In the Standard Model of particle physics, electrons belong to the group of subatomic particles called leptons, which are believed to be fundamental or elementary particles. Electrons have the lowest mass of any charged lepton (or electrically charged particle of any type) and belong to the first-generation of fundamental particles. The second and third generation contain charged leptons, the muon and the tau, which are identical to the electron in charge, spin and interactions, but are more massive. Leptons differ from the other basic constituent of matter, the quarks, by their lack of strong interaction. All members of the lepton group are fermions, because they all have half-odd integer spin; the electron has spin $\frac{1}{2}$.

Fundamental Properties

The invariant mass of an electron is approximately 9.109×10^{-31} kilograms, or 5.489×10^{-4} atomic mass units. On the basis of Einstein's principle of mass–energy equivalence, this mass corresponds to a rest energy of 0.511 MeV. The ratio between the mass of a proton and that of an electron is about 1836. Astronomical measurements show that the proton-to-electron mass ratio has held the same value, as is predicted by the Standard Model, for at least half the age of the universe.

Electrons have an electric charge of -1.602×10^{-19} coulombs, which is used as a standard unit of charge for subatomic particles, and is also called the elementary charge. This elementary charge has a relative standard uncertainty of 2.2×10^{-8}. Within the limits of experimental accuracy, the electron charge is identical to the charge of a proton, but with the opposite sign. As the symbol e is used for the elementary charge, the electron is commonly symbolized by e^-, where the minus sign indicates the negative charge. The positron is symbolized by e^+ because it has the same properties as the electron but with a positive rather than negative charge.

The electron has an intrinsic angular momentum or spin of $\frac{1}{2}$. This property is usually stated by referring to the electron as a spin $-\frac{1}{2}$ particle. For such particles the spin magnitude is $\frac{\sqrt{3}}{2}\hbar$. while the result of the measurement of a projection of the spin on any axis can only be $\pm\frac{\hbar}{2}$. In addition to spin, the electron has an intrinsic magnetic moment along its spin axis. It is approximately equal to one Bohr magneton, which is a physical constant equal to $9.27400915(23) \times 10^{-24}$ joules per tesla. The orientation of the spin with respect to the momentum of the electron defines the property of elementary particles known as helicity.

The electron has no known substructure and it is assumed to be a point particle with a point charge and no spatial extent.

The issue of the radius of the electron is a challenging problem of the modern theoretical physics. The admission of the hypothesis of a finite radius of the electron is incompatible to the premises of the theory of relativity. On the other hand, a point-like electron (zero radius) generates serious mathematical difficulties due to the self-energy of the electron tending to infinity. Observation of a single electron in a Penning trap suggests the upper limit of the particle's radius to be 10^{-22} meters. The upper bound of the electron radius of 10^{-18} meters can be derived using the uncertainty relation in energy. There is also a physical constant called the "classical electron radius", with the much larger value of 2.8179×10^{-15} m, greater than the radius of the proton. However, the terminology comes from a simplistic calculation that ignores the effects of quantum mechanics; in reality, the so-called classical electron radius has little to do with the true fundamental structure of the electron.

There are elementary particles that spontaneously decay into less massive particles. An example is the muon, with a mean lifetime of 2.2×10^{-6} seconds, which decays into an electron, a muon neutrino and an electron antineutrino. The electron, on the other hand, is thought to be stable on theoretical grounds: the electron is the least massive particle with non-zero electric charge, so its decay would violate charge conservation. The experimental lower bound for the electron's mean lifetime is 6.6×10^{28} years, at a 90% confidence level.

Quantum Properties

As with all particles, electrons can act as waves. This is called the wave–particle duality and can be demonstrated using the double-slit experiment.

The wave-like nature of the electron allows it to pass through two parallel slits simultaneously, rather than just one slit as would be the case for a classical particle. In quantum mechanics, the wave-like property of one particle can be described mathematically as a complex-valued function, the wave function, commonly denoted by the Greek letter psi (ψ). When the absolute value of this function is squared, it gives the probability that a particle will be observed near a location—a probability density.

Electrons are identical particles because they cannot be distinguished from each other by their intrinsic physical properties. In quantum mechanics, this means that a pair of interacting electrons must be able to swap positions without an observable change to the state of the system. The wave function of fermions, including electrons, is antisymmetric, meaning that it changes sign when two electrons are swapped; that is, $\psi\left(r_1, r_2\right) = -\psi\left(r_2, r_1\right)$, where the variables r_1 and r_2 correspond to the first and second electrons, respectively. Since the absolute value is not changed by a sign swap, this corresponds to equal probabilities. Bosons, such as the photon, have symmetric wave functions instead.

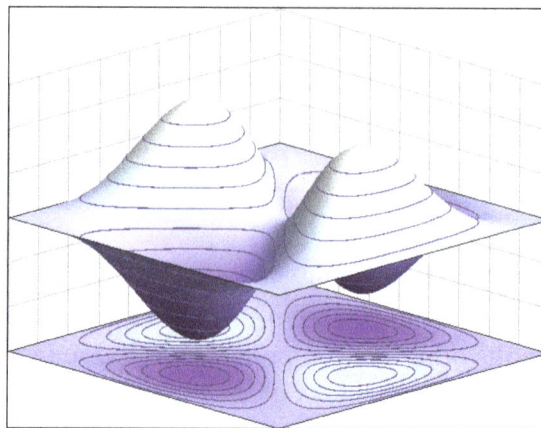

Example of an antisymmetric wave function for a quantum state of two identical fermions in a 1-dimensional box. If the particles swap position, the wave function inverts its sign.

In the case of antisymmetry, solutions of the wave equation for interacting electrons result in a zero probability that each pair will occupy the same location or state. This is responsible for the Pauli exclusion principle, which precludes any two electrons from occupying the same quantum state. This principle explains many of the properties of electrons. For example, it causes groups of bound electrons to occupy different orbitals in an atom, rather than all overlapping each other in the same orbit.

Virtual Particles

In a simplified picture, every photon spends some time as a combination of a virtual electron plus its antiparticle, the virtual positron, which rapidly annihilate each other shortly thereafter. The combination of the energy variation needed to create these particles, and the time during which they exist, fall under the threshold of detectability expressed by the Heisenberg uncertainty relation, $\Delta E \cdot \Delta t \geq \hbar$. In effect, the energy needed to create these virtual particles, ΔE, can be "borrowed" from the vacuum for a period of time, Δt, so that their product is no more than the reduced Planck constant, $\hbar \approx 6.6 \times 10^{-16}\ eV \cdot s$. Thus, for a virtual electron, Δt is at most 1.3×10^{-21} s.

While an electron–positron virtual pair is in existence, the coulomb force from the ambient electric field surrounding an electron causes a created positron to be attracted to the original electron, while a created electron experiences a repulsion. This causes what is called vacuum polarization. In effect, the vacuum behaves like a medium having a dielectric permittivity more than unity. Thus the effective charge of an electron is actually smaller than its true value, and the charge decreases with increasing distance from the electron. This polarization was confirmed experimentally in 1997 using the Japanese TRISTAN particle accelerator. Virtual particles cause a comparable shielding effect for the mass of the electron.

The interaction with virtual particles also explains the small (about 0.1%) deviation of the intrinsic magnetic moment of the electron from the Bohr magneton (the anomalous magnetic moment). The extraordinarily precise agreement of this predicted difference with the experimentally determined value is viewed as one of the great achievements of quantum electrodynamics.

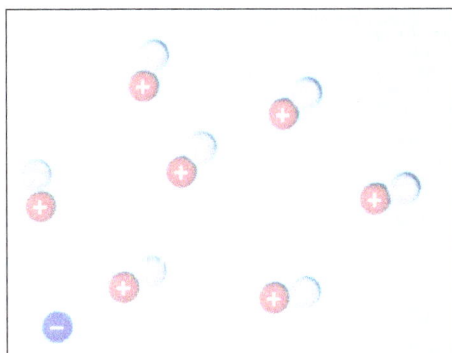

A schematic depiction of virtual electron–positron pairs appearing at random near an electron (at lower left).

The apparent paradox in classical physics of a point particle electron having intrinsic angular momentum and magnetic moment can be explained by the formation of virtual photons in the electric field generated by the electron. These photons cause the electron to shift about in a jittery fashion (known as zitterbewegung), which results in a net circular motion with precession. This motion produces both the spin and the magnetic moment of the electron. In atoms, this creation of virtual photons explains the Lamb shift observed in spectral lines.

Interaction

An electron generates an electric field that exerts an attractive force on a particle with a positive charge, such as the proton, and a repulsive force on a particle with a negative charge. The strength of this force in nonrelativistic approximation is determined by Coulomb's inverse square law. When an electron is in motion, it generates a magnetic field. The Ampère-Maxwell law relates the magnetic field to the mass motion of electrons (the current) with respect to an observer. This property of induction supplies the magnetic field that drives an electric motor. The electromagnetic field of an arbitrary moving charged particle is expressed by the Liénard–Wiechert potentials, which are valid even when the particle's speed is close to that of light (relativistic).

When an electron is moving through a magnetic field, it is subject to the Lorentz force that acts perpendicularly to the plane defined by the magnetic field and the electron velocity. This centripetal force causes the electron to follow a helical trajectory through the field at a radius called the gyroradius. The acceleration from this curving motion induces the electron to radiate energy in the form of synchrotron radiation. The energy emission in turn causes a recoil of the electron, known as the Abraham–Lorentz–Dirac Force, which creates a friction that slows the electron. This force is caused by a back-reaction of the electron's own field upon itself.

Photons mediate electromagnetic interactions between particles in quantum electrodynamics. An isolated electron at a constant velocity cannot emit or absorb a real photon; doing so would violate conservation of energy and momentum. Instead, virtual photons can transfer momentum between two charged particles. This exchange of virtual photons, for example, generates the Coulomb force. Energy emission can occur when a moving electron is deflected by a charged particle, such as a proton. The acceleration of the electron results in the emission of Bremsstrahlung radiation.

An inelastic collision between a photon (light) and a solitary (free) electron is called Compton scattering. This collision results in a transfer of momentum and energy between the particles, which modifies the wavelength of the photon by an amount called the Compton shift. The maximum magnitude of this wavelength shift is h/mec, which is known as the Compton wavelength. For an electron, it has a value of 2.43×10^{-12} m. When the wavelength of the light is long (for instance, the wavelength of the visible light is $0.4 - 0.7$ μm) the wavelength shift becomes negligible. Such interaction between the light and free electrons is called Thomson scattering or linear Thomson scattering.

The relative strength of the electromagnetic interaction between two charged particles, such as an electron and a proton, is given by the fine-structure constant. This value is a dimensionless quantity formed by the ratio of two energies: the electrostatic energy of attraction (or repulsion) at a separation of one Compton wavelength, and the rest energy of the charge. It is given by $\alpha \approx 7.297353 \times 10^{-3}$, which is approximately equal to $\frac{1}{137}$.

When electrons and positrons collide, they annihilate each other, giving rise to two or more gamma ray photons. If the electron and positron have negligible momentum, a positronium atom can form before annihilation results in two or three gamma ray photons totalling 1.022 MeV. On the other hand, a high-energy photon can transform into an electron and a positron by a process called pair production, but only in the presence of a nearby charged particle, such as a nucleus.

In the theory of electroweak interaction, the left-handed component of electron's wavefunction forms a weak isospin doublet with the electron neutrino. This means that during weak interactions, electron neutrinos behave like electrons. Either member of this doublet can undergo a charged current interaction by emitting or absorbing a W and be converted into the other member. Charge is conserved during this reaction because the W boson also carries a charge, canceling out any net change during the transmutation. Charged current interactions are responsible for the phenomenon of beta decay in a radioactive atom. Both the electron and electron neutrino can undergo a neutral current interaction via a Z^o exchange, and this is responsible for neutrino-electron elastic scattering.

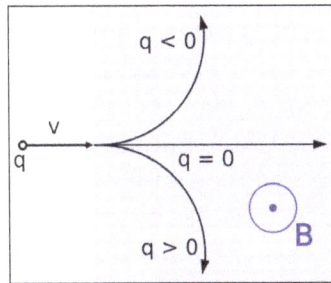

A particle with charge q (at left) is moving with velocity v through a magnetic field B that is oriented toward the viewer. For an electron, q is negative so it follows a curved trajectory toward the top.

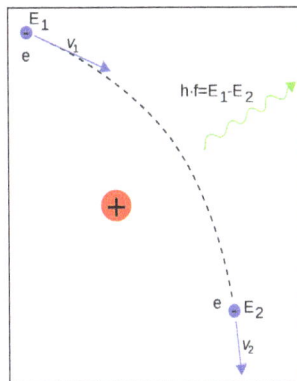

Here, Bremsstrahlung is produced by an electron e deflected by the electric field of an atomic nucleus. The energy change $E_2 - E_1$ determines the frequency f of the emitted photon.

Atoms and Molecules

An electron can be bound to the nucleus of an atom by the attractive Coulomb force. A system of one or more electrons bound to a nucleus is called an atom. If the number of electrons is different from the nucleus' electrical charge, such an atom is called an ion. The wave-like behavior of a bound electron is described by a function called an atomic orbital. Each orbital has its own set of quantum numbers such as energy, angular momentum and projection of angular momentum, and only a discrete set of these orbitals exist around the nucleus. According to the Pauli exclusion principle each orbital can be occupied by up to two electrons, which must differ in their spin quantum number.

Electrons can transfer between different orbitals by the emission or absorption of photons with an energy that matches the difference in potential. Other methods of orbital transfer include collisions with particles, such as electrons, and the Auger effect. To escape the atom, the energy of the electron must be increased above its binding energy to the atom. This occurs, for example, with the photoelectric effect, where an incident photon exceeding the atom's ionization energy is absorbed by the electron.

The orbital angular momentum of electrons is quantized. Because the electron is charged, it produces an orbital magnetic moment that is proportional to the angular momentum. The net magnetic moment of an atom is equal to the vector sum of orbital and spin magnetic moments of all electrons and the nucleus. The magnetic moment of the nucleus is negligible compared with that of the electrons. The magnetic moments of the electrons that occupy the same orbital (so called, paired electrons) cancel each other out.

The chemical bond between atoms occurs as a result of electromagnetic interactions, as described by the laws of quantum mechanics. The strongest bonds are formed by the sharing or transfer of electrons between atoms, allowing the formation of molecules. Within a molecule, electrons move under the influence of several nuclei, and occupy molecular orbitals; much as they can occupy atomic orbitals in isolated atoms. A fundamental factor in these molecular structures is the existence of electron pairs. These are electrons with opposed spins, allowing them to occupy the same molecular orbital without violating the Pauli exclusion principle (much like in atoms). Different molecular orbitals have different spatial distribution of the electron density. For instance, in bonded pairs (i.e. in the pairs that actually bind atoms together) electrons can be found with the maximal probability in a relatively small volume between the nuclei. By contrast, in non-bonded pairs electrons are distributed in a large volume around nuclei.

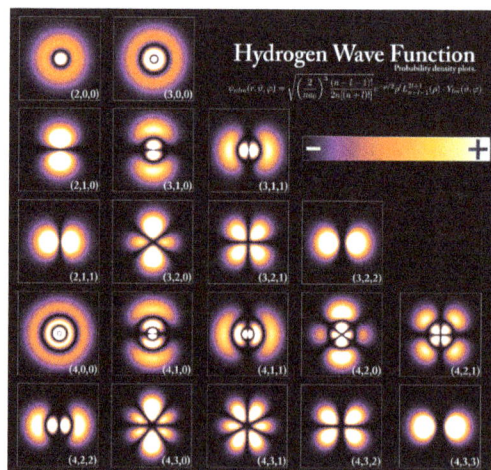

Probability densities for the first few hydrogen atom orbitals, seen in cross-section. The energy level of a bound electron determines the orbital it occupies, and the color reflects the probability of finding the electron at a given position.

Conductivity

If a body has more or fewer electrons than are required to balance the positive charge of the nuclei, then that object has a net electric charge. When there is an excess of electrons, the object is said to be negatively charged. When there are fewer electrons than the number of protons in nuclei, the object is said to be positively charged. When the number of electrons and the number of protons are equal, their charges cancel each other and the object is said to be electrically neutral. A macroscopic body can develop an electric charge through rubbing, by the triboelectric effect.

Independent electrons moving in vacuum are termed free electrons. Electrons in metals also behave as if they were free. In reality the particles that are commonly termed electrons in metals and other solids are quasi-electrons—quasiparticles, which have the same electrical charge, spin, and magnetic moment as real electrons but might have a different mass. When free electrons—both in vacuum and metals—move, they produce a net flow of charge called an electric current, which generates a magnetic field. Likewise a current can be created by a changing magnetic field. These interactions are described mathematically by Maxwell's equations.

At a given temperature, each material has an electrical conductivity that determines the value of electric current when an electric potential is applied. Examples of good conductors include metals such as copper and gold, whereas glass and Teflon are poor conductors. In any dielectric material, the electrons remain bound to their respective atoms and the material behaves as an insulator. Most semiconductors have a variable level of conductivity that lies between the extremes of conduction and insulation. On the other hand, metals have an electronic band structure containing partially filled electronic bands. The presence of such bands allows electrons in metals to behave as if they were free or delocalized electrons. These electrons are not associated with specific atoms, so when an electric field is applied, they are free to move like a gas (called Fermi gas) through the material much like free electrons.

Because of collisions between electrons and atoms, the drift velocity of electrons in a conductor is on the order of millimeters per second. However, the speed at which a change of current at one point in the material causes changes in currents in other parts of the material, the velocity of propagation, is typically about 75% of light speed. This occurs because electrical signals propagate as a wave, with the velocity dependent on the dielectric constant of the material.

Metals make relatively good conductors of heat, primarily because the delocalized electrons are free to transport thermal energy between atoms. However, unlike electrical conductivity, the thermal conductivity of a metal is nearly independent of temperature. This is expressed mathematically by the Wiedemann–Franz law, which states that the ratio of thermal conductivity to the electrical conductivity is proportional to the temperature. The thermal disorder in the metallic lattice increases the electrical resistivity of the material, producing a temperature dependence for electric current.

A lightning discharge consists primarily of a flow of electrons. The electric potential needed for lightning can be generated by a triboelectric effect.

When cooled below a point called the critical temperature, materials can undergo a phase transition in which they lose all resistivity to electric current, in a process known as superconductivity. In BCS theory, pairs of electrons called Cooper pairs have their motion coupled to nearby matter via lattice vibrations called phonons, thereby avoiding the collisions with atoms that normally create electrical resistance. (Cooper pairs have a radius of roughly 100 nm, so they can overlap each other.) However, the mechanism by which higher temperature superconductors operate remains uncertain.

Electrons inside conducting solids, which are quasi-particles themselves, when tightly confined at temperatures close to absolute zero, behave as though they had split into three other quasiparticles: spinons, orbitons and holons. The former carries spin and magnetic moment, the next carries its orbital location while the latter electrical charge.

Motion and Energy

According to Einstein's theory of special relativity, as an electron's speed approaches the speed of light, from an observer's point of view its relativistic mass increases, thereby making it more and more difficult to accelerate it from within the observer's frame of reference. The speed of an electron can approach, but never reach, the speed of light in a vacuum, c. However, when relativistic electrons—that is, electrons moving at a speed close to c—are injected into a dielectric medium such as water, where the local speed of light is significantly less than c, the electrons temporarily travel faster than light in the medium. As they interact with the medium, they generate a faint light called Cherenkov radiation.

The effects of special relativity are based on a quantity known as the Lorentz factor, defined as $\gamma = 1/\sqrt{1 - v^2/c^2}$ particle. The kinetic energy K_e of an electron moving with velocity v is:

$$K_e = (\gamma - 1)m_e c^2,$$

where m_e is the mass of electron. For example, the Stanford linear accelerator can accelerate an electron to roughly 51 GeV. Since an electron behaves as a wave, at a given velocity it has a characteristic de Broglie wavelength. This is given by $\lambda e = h/p$ where h is the Planck constant and p is the momentum. For the 51 GeV electron above, the wavelength is about 2.4×10^{-17} m, small enough to explore structures well below the size of an atomic nucleus.

Formation

The Big Bang theory is the most widely accepted scientific theory to explain the early stages in the evolution of the Universe. For the first millisecond of the Big Bang, the temperatures were over 10 billion kelvins and photons had mean energies over a million electronvolts. These photons were sufficiently energetic that they could react with each other to form pairs of electrons and positrons. Likewise, positron-electron pairs annihilated each other and emitted energetic photons:

$$\gamma + \gamma \leftrightarrow e^+ + e^-$$

An equilibrium between electrons, positrons and photons was maintained during this phase of the evolution of the Universe. After 15 seconds had passed, however, the temperature of the universe dropped below the threshold where electron-positron formation could occur. Most of the surviving electrons and positrons annihilated each other, releasing gamma radiation that briefly reheated the universe.

For reasons that remain uncertain, during the annihilation process there was an excess in the number of particles over antiparticles. Hence, about one electron for every billion electron-positron pairs survived. This excess matched the excess of protons over antiprotons, in a condition known as baryon asymmetry, resulting in a net charge of zero for the universe. The surviving protons and neutrons began to participate in reactions with each other—in the process known as nucleosynthesis, forming isotopes of hydrogen and helium, with trace amounts of lithium. This process peaked after about five minutes. Any leftover neutrons underwent negative beta decay with a half-life of about a thousand seconds, releasing a proton and electron in the process,

$$n \rightarrow p + e^- + \bar{v}_e$$

For about the next $300000 - 400000$ years, the excess electrons remained too energetic to bind with atomic nuclei. What followed is a period known as recombination, when neutral atoms were formed and the expanding universe became transparent to radiation.

Roughly one million years after the big bang, the first generation of stars began to form. Within a star, stellar nucleosynthesis results in the production of positrons from the fusion of atomic nuclei. These antimatter particles immediately annihilate with electrons, releasing gamma rays. The net result is a steady reduction in the number of electrons, and a matching increase in the number of neutrons. However, the process of stellar evolution can result in the synthesis of radioactive isotopes. Selected isotopes can subsequently undergo negative beta decay, emitting an electron and antineutrino from the nucleus. An example is the cobalt-60 (^{60}Co) isotope, which decays to form nickel-60 (^{60}Ni).

At the end of its lifetime, a star with more than about 20 solar masses can undergo gravitational collapse to form a black hole. According to classical physics, these massive stellar objects exert a gravitational attraction that is strong enough to prevent anything, even electromagnetic radiation, from escaping past the Schwarzschild radius. However, quantum mechanical effects are believed to potentially allow the emission of Hawking radiation at this distance. Electrons (and positrons) are thought to be created at the event horizon of these stellar remnants.

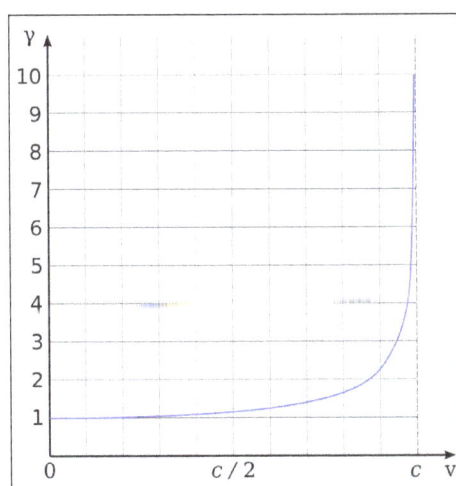

Lorentz factor as a function of velocity. It starts at value 1 and goes to infinity as v approaches c.

When a pair of virtual particles (such as an electron and positron) is created in the vicinity of the event horizon, random spatial positioning might result in one of them to appear on the exterior;

this process is called quantum tunnelling. The gravitational potential of the black hole can then supply the energy that transforms this virtual particle into a real particle, allowing it to radiate away into space. In exchange, the other member of the pair is given negative energy, which results in a net loss of mass-energy by the black hole. The rate of Hawking radiation increases with decreasing mass, eventually causing the black hole to evaporate away until, finally, it explodes.

Cosmic rays are particles traveling through space with high energies. Energy events as high as 3.0×10^{20} eV have been recorded. When these particles collide with nucleons in the Earth's atmosphere, a shower of particles is generated, including pions. More than half of the cosmic radiation observed from the Earth's surface consists of muons. The particle called a muon is a lepton produced in the upper atmosphere by the decay of a pion.

$$\pi^- \rightarrow \mu^- + \bar{\nu}_\mu$$

A muon, in turn, can decay to form an electron or positron.

$$\mu^- \rightarrow e^- + \bar{\nu}_e + \nu_\mu$$

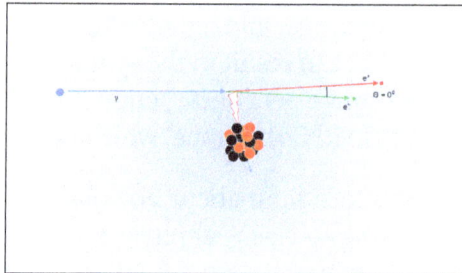

Pair production of an electron and positron, caused by the close approach of a photon with an atomic nucleus. The lightning symbol represents an exchange of a virtual photon, thus an electric force acts. The angle between the particles is very small.

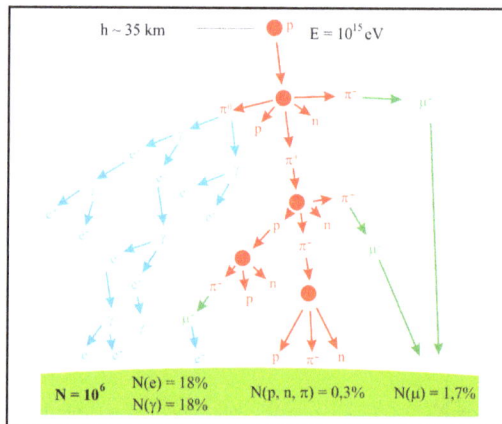

An extended air shower generated by an energetic cosmic ray striking the Earth's atmosphere.

Observation

Remote observation of electrons requires detection of their radiated energy. For example, in high-energy environments such as the corona of a star, free electrons form a plasma that radiates energy due to Bremsstrahlung radiation. Electron gas can undergo plasma oscillation, which is waves caused by synchronized variations in electron density, and these produce energy emissions that can be detected by using radio telescopes.

The frequency of a photon is proportional to its energy. As a bound electron transitions between different energy levels of an atom, it absorbs or emits photons at characteristic frequencies. For instance, when atoms are irradiated by a source with a broad spectrum, distinct absorption lines appear in the spectrum of transmitted radiation. Each element or molecule displays a characteristic set of spectral lines, such as the hydrogen spectral series. Spectroscopic measurements of the strength and width of these lines allow the composition and physical properties of a substance to be determined.

In laboratory conditions, the interactions of individual electrons can be observed by means of particle detectors, which allow measurement of specific properties such as energy, spin and charge. The development of the Paul trap and Penning trap allows charged particles to be contained within a small region for long durations. This enables precise measurements of the particle properties. For example, in one instance a Penning trap was used to contain a single electron for a period of 10 months. The magnetic moment of the electron was measured to a precision of eleven digits, which, in 1980, was a greater accuracy than for any other physical constant.

The first video images of an electron's energy distribution were captured by a team at Lund University in Sweden, February 2008. The scientists used extremely short flashes of light, called attosecond pulses, which allowed an electron's motion to be observed for the first time.

The distribution of the electrons in solid materials can be visualized by angle-resolved photoemission spectroscopy (ARPES). This technique employs the photoelectric effect to measure the reciprocal space—a mathematical representation of periodic structures that is used to infer the original structure. ARPES can be used to determine the direction, speed and scattering of electrons within the material.

Aurorae are mostly caused by energetic electrons precipitating into the atmosphere.

Plasma Applications

Particle Beams

Electron beams are used in welding. They allow energy densities up to 10^7 W·cm^{-2} across a narrow focus diameter of 0.1 – 1.3 mm and usually require no filler material. This welding technique must be performed in a vacuum to prevent the electrons from interacting with the gas before reaching their target, and it can be used to join conductive materials that would otherwise be considered unsuitable for welding.

Electron-beam lithography (EBL) is a method of etching semiconductors at resolutions smaller than a micrometer. This technique is limited by high costs, slow performance, the need to operate the beam in the vacuum and the tendency of the electrons to scatter in solids. The last problem limits the resolution to about 10 nm. For this reason, EBL is primarily used for the production of small numbers of specialized integrated circuits.

Electron beam processing is used to irradiate materials in order to change their physical properties or sterilize medical and food products. Electron beams fluidise or quasi-melt glasses without significant increase of temperature on intensive irradiation: e.g., intensive electron radiation causes a many orders of magnitude decrease of viscosity and stepwise decrease of its activation energy.

Linear particle accelerators generate electron beams for treatment of superficial tumors in radiation therapy. Electron therapy can treat such skin lesions as basal-cell carcinomas because an electron beam only penetrates to a limited depth before being absorbed, typically up to 5 cm for electron energies in the range 5–20 MeV. An electron beam can be used to supplement the treatment of areas that have been irradiated by X-rays.

Particle accelerators use electric fields to propel electrons and their antiparticles to high energies. These particles emit synchrotron radiation as they pass through magnetic fields. The dependency of the intensity of this radiation upon spin polarizes the electron beam—a process known as the Sokolov–Ternov effect. Polarized electron beams can be useful for various experiments. Synchrotron radiation can also cool the electron beams to reduce the momentum spread of the particles. Electron and positron beams are collided upon the particles' accelerating to the required energies; particle detectors observe the resulting energy emissions, which particle physics studies .

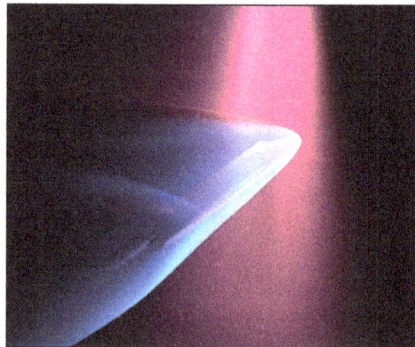

Imaging

Low-energy electron diffraction (LEED) is a method of bombarding a crystalline material with a collimated beam of electrons and then observing the resulting diffraction patterns to determine the structure of the material. The required energy of the electrons is typically in the range 20–200 eV. The reflection high-energy electron diffraction (RHEED) technique uses the reflection of a beam of electrons fired at various low angles to characterize the surface of crystalline materials. The beam energy is typically in the range 8–20 keV and the angle of incidence is 1–4°.

The electron microscope directs a focused beam of electrons at a specimen. Some electrons change their properties, such as movement direction, angle, and relative phase and energy as the beam interacts with the material. Microscopists can record these changes in the electron beam to produce atomically resolved images of the material. In blue light, conventional optical microscopes have a diffraction-limited resolution of about 200 nm. By comparison, electron microscopes are limited by the de Broglie wavelength of the electron. This wavelength, for example, is equal to 0.0037 nm for electrons accelerated across a 100,000-volt potential. The Transmission Electron Aberration-Corrected Microscope is capable of sub-0.05 nm resolution, which is more than enough to resolve individual atoms. This capability makes the electron microscope a useful laboratory instrument for high resolution imaging. However, electron microscopes are expensive instruments that are costly to maintain.

Two main types of electron microscopes exist: transmission and scanning. Transmission electron microscopes function like overhead projectors, with a beam of electrons passing through a slice of material then being projected by lenses on a photographic slide or a charge-coupled device. Scanning electron microscopes rasteri a finely focused electron beam, as in a TV set, across the studied sample to produce the image. Magnifications range from $100\times$ *to* $1,000,000\times$ or higher for both microscope types. The scanning tunneling microscope uses quantum tunneling of electrons from a sharp metal tip into the studied material and can produce atomically resolved images of its surface.

Other Applications

In the free-electron laser (FEL), a relativistic electron beam passes through a pair of undulators that contain arrays of dipole magnets whose fields point in alternating directions. The electrons emit synchrotron radiation that coherently interacts with the same electrons to strongly amplify the radiation field at the resonance frequency. FEL can emit a coherent high-brilliance electromagnetic radiation with a wide range of frequencies, from microwaves to soft X-rays. These devices are used in manufacturing, communication, and in medical applications, such as soft tissue surgery.

Electrons are important in cathode ray tubes, which have been extensively used as display devices in laboratory instruments, computer monitors and television sets. In a photomultiplier tube, every photon striking the photocathode initiates an avalanche of electrons that produces a detectable current pulse. Vacuum tubes use the flow of electrons to manipulate electrical signals, and they played a critical role in the development of electronics technology. However, they have been largely supplanted by solid-state devices such as the transistor.

The Charge to Mass Ratio of an Electron

The history of the atomic structure and quantum mechanics dates back to the times of Democritus, the man who first proposed that matter is composed of atoms. These theories could not gain much importance due to the lack of technology. The experiments conducted during the nineteenth century and early twentieth century revealed that even atom is not the ultimate particle. The continued efforts of the scientists led to the discovery of subatomic particles like electrons, protons, and neutrons.

In the nineteenth century, J.J Thomson proposed Thomson's Atomic Model discovered the electron to mark an inception to the world of subatomic particles. Once the electron was discovered, he continued his experiments to calculate the charge and the mass of the electron. With the help of his experiments, he derived a formula for the calculation of charge to mass ratio of the electron.

The charge to mass ratio of the electron is given by :

$$e/m \;=\; 1.758820 \times 10^{11}\,C/kg$$

Where,

m = mass of an electron in kg = $9.10938356 \times 10^{-31}$ kilograms.

e = magnitude of the charge of an electron in coulombs = $1.602\; x\; 10^{-19}$ coulombs.

Experimental Setup for the Determination of Charge to Mass Ratio of Electron

While carrying out discharge tube experiment, Thomson observed that the particles of the cathode deviate from their path. He noticed the amount of deviation in the presence of electrical or magnetic field depends on various related parameters. They are:

- Particles with a greater magnitude of the charge experienced greater interaction with the electric or magnetic field. Thus, they exhibited a greater deflection.

- Lighter particle experienced greater deflection. Thus, deflection is inversely proportional to the mass of the particle.

- Deflection of the particle from their path is directly proportional to the strength of the electrical and the magnetic field present.

Let us now understand this with the help of his experimental observations:

- The electrons deviated from their path and hit the cathode ray tube at point 'x' in the presence of a lone electric field.

- Similarly, electron strikes the discharge tube at point 'z' when only the magnetic field was present.

- Thus, to make electrons continue on the same path we need to balance the electric and magnetic field acting on them.

- Finally, based on the deflection of the electron, Thomson calculated the value of charge to mass ratio of the electron.

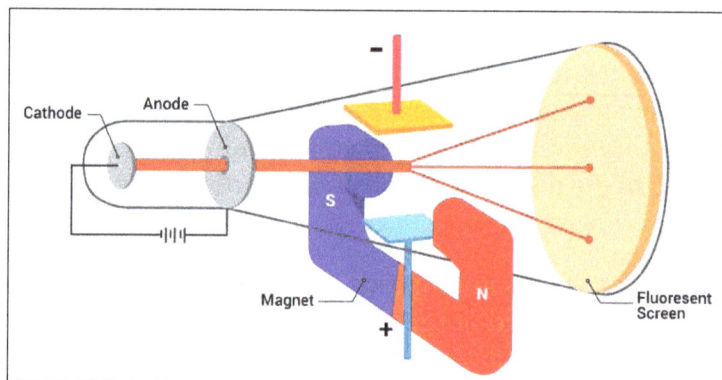

ATOMIC UNITS

Atomic units (au or a.u.) form a system of natural units which is especially convenient for atomic physics calculations. There are two different kinds of atomic units, Hartree atomic units and Rydberg atomic units, which differ in the choice of the unit of mass and charge. This topic deals with Hartree atomic units, where the numerical values of the following four fundamental physical constants are all unity by definition:

- Electron mass m_e;

- Elementary charge e;

- Reduced Planck's constant $\hbar = h/(2\pi)$;

- Coulomb's constant $k_e = 1/(4\pi\varepsilon_0)$.

In Hartree units, the speed of light is approximately 137. Atomic units are often abbreviated "a.u." or "au", used also for astronomical units, arbitrary units, and absorbance units in different contexts.

Use and Notation

Atomic units, like SI units, have a unit of mass, a unit of length, and so on. However, the use and notation is somewhat different from SI.

Suppose a particle with a mass of m has 3.4 times the mass of electron. The value of m can be written in three ways:

- "$m = 3.4m_e$". This is the clearest notation (but least common), where the atomic unit is included explicitly as a symbol.

- '$m = 3.4\,a.u.$" ("a.u." means "expressed in atomic units"). This notation is ambiguous: Here, it means that the mass m is 3.4 times the atomic unit of mass. But if a length L were 3.4 times the atomic unit of length, the equation would look the same, '$L = 3.4\ a.u$" The dimension needs to be inferred from context.

- '$m = 3.4$". This notation is similar to the previous one, and has the same dimensional ambiguity. It comes from formally setting the atomic units to 1, in this case $m_e = 1$, so $3.4\ m_e = 3.4$.

Fundamental Atomic Units

These four fundamental constants form the basis of the atomic units. Therefore, their numerical values in the atomic units are unity by definition.

Dimension	Name	Symbol/Definition	Value in SI units
mass	electron rest mass	m_e	$9.1093837015(28) \times 10^{-31}\ kg$

charge	elementary charge	e	$1.602176634(0) \times 10^{-19} \, C$
Action	reduced Planck's constant	$\hbar = h/(2\pi)$	$1.054571817(0) \times 10^{-34} \, J{\cdot}s$
electric constant^{-1}	Coulomb force constant	$k_e = 1/(4\pi\epsilon_0)$	$8.987551792(1) \times 10^9 \, kg{\cdot}m^3{\cdot}s^{-2}{\cdot}C^{-2}$

Related Physical Constants

Dimensionless physical constants retain their values in any system of units. Of particular importance is the fine-structure constant $\alpha = \dfrac{e^2}{(4\pi\epsilon_0)\hbar c} \approx 1/137$. This immediately gives the value of the speed of light, expressed in atomic units.

Some Physical Constants Expressed in Atomic Units

Name	Symbol/Definition	Value in atomic units
speed of light	c	$1/\alpha \approx 137$
classical electron radius	$r_e = \dfrac{1}{4\pi\epsilon_0}\dfrac{e^2}{m_e c^2}$	$\alpha^2 \approx 5.32 \times 10^{-5}$
proton mass	m_p	$m_p/m_e \approx 1836$

Derived Atomic Units

Below are given a few derived units. Some of them have proper names and symbols assigned, as indicated in the table. k_B is the Boltzmann constant.

Dimension	Name	Symbol	Expression	Value in SI units	Value in other units
length	bohr	a_0	$4\pi\epsilon_0\hbar^2/(m_e e^2) = \hbar/(m_e c\alpha)$	$5.29177210903(80) \times 10^{-11} \, m$	$0.0529177210903(80) \, nm =$ $0.529177210903(80) \, \mathring{A}$
energy	hartree	E_h	$m_e e^4/(4\pi\epsilon_0\hbar)^2 = \alpha^2 m_e c^2$	$4.3597447222071(85) \times 10^{-18} \, J$	$27.211386245988 \, eV =$ $627.509 \, kcal{\cdot}mol^{-1}$
time			$\hbar/E_h$	$2.4188843265857(47) \times 10^{-17} \, s$	
momentum			$\hbar/a_0$	$1.99285191410(30) \times 10^{-24} \, kg{\cdot}m{\cdot}s^{-1}$	
velocity			$a_0 E_h/\hbar = \alpha c$	$2.18769126364(33) \times 10^6 \, m{\cdot}s^{-1}$	
force			E_h/a_0	$8.2387234983(12) \times 10^{-8} \, N$	$82.387 \, nN = 51.421 \, eV{\cdot}\mathring{A}^{-1}$
temperature			E_h/k_B	$3.1577464(55) \times 10^5 \, K$	
pressure			E_h/a_0^3	$2.9421912(19) \times 10^{13} \, Pa$	

electric field			$E_h/(ea_0)$	$5.14220674763(78)\times10^{11}\,V\cdot m^{-1}$	$5.14220674763(78)\,GV\cdot cm^{-1} =$ $51.4220674763(78)\,V\cdot \text{Å}^{-1}$
electric potential			E_h/e	$2.7211386245988(53)\times10^1\,V$	$2.541746473\,D$
electric dipole moment			ea_0	$8.4783536255(13)\times10^{-30}\,C\cdot m$	$2.35051756758(71)\times10^9\,G$
Magnetic field (SI)			$\dfrac{\rule{1em}{0.4pt}}{ea}$	$2.35051756758(71)\times10^5\,T$	2 (Bohr magneton)
Magnetic dipole moment (SI)			$\dfrac{e\hbar}{m_e}$		$1.715255528(11)\times10^7\,G$
Magnetic field (cgs)			$\dfrac{e}{a_0^2 c}$	$1.715255528(11)\times10^3\,T$	2 (Bohr magneton)
Magnetic dipole moment (cgs)			$\dfrac{e\hbar}{m_e c}$		

SI and Gaussian-CGS Variants and Magnetism-related Units

There are two common variants of atomic units, one where they are used in conjunction with SI units for electromagnetism, and one where they are used with Gaussian-CGS units. Although most of the units listed above are the same either way (including the unit for electric field), the units related to magnetism are not. In the SI system, the atomic unit for magnetic field is:

$$1 \text{ a.u.} = \frac{\hbar}{ea_0^2} \approx 2.35\times10^5\text{ T} = 2.35\times10^9\text{ G},$$

and in the Gaussian-cgs unit system, the atomic unit for magnetic field is:

$$1 \text{ a.u.} = \frac{e}{a_0^2 c} \approx 1.72\times10^3\text{ T} = 1.72\times10^7\text{ G}.$$

(These differ by a factor of α.)

Other magnetism-related quantities are also different in the two systems. An important example is the Bohr magneton: In SI-based atomic units,

$$\mu_B = \frac{e\hbar}{2m_e} = 1/2 \ a.u.$$

and in Gaussian-based atomic units,

$$\mu_B = \frac{e\hbar}{2m_e c} = \alpha/2 \approx 3.6\times10^{-3}\,a.u.$$

Bohr Model in Atomic Units

Atomic units are chosen to reflect the properties of electrons in atoms. This is particularly clear from the classical Bohr model of the hydrogen atom in its ground state. The ground state electron orbiting the hydrogen nucleus has (in the classical Bohr model):

- Orbital velocity = 1

- Orbital radius = 1

- Angular momentum = 1

- Orbital period = 2 π

- Ionization energy = ½

- Electric field (due to nucleus) = 1

- Electrical attractive force (due to nucleus) = 1

Non-relativistic Quantum Mechanics in Atomic Units

The Schrödinger equation for an electron in SI units is:

$$-\frac{\hbar^2}{2m_e}\nabla^2\psi(r,t)+V(r)\psi(r,t)=i\hbar\frac{\partial\psi}{\partial t}(r,t).$$

For the special case of the electron around a hydrogen atom, the Hamiltonian in SI units is:

$$\hat{H}=-\frac{\hbar^2}{2m_e}\nabla^2-\frac{1}{4\pi\epsilon_0}\frac{e^2}{r},$$

while atomic units transform the preceding equation into:

$$\hat{H}=-\frac{1}{2}\nabla^2-\frac{1}{r}.$$

Comparison with Planck Units

Both Planck units and au are derived from certain fundamental properties of the physical world, and are free of anthropocentric considerations. It should be kept in mind that au were designed for atomic-scale calculations in the present-day universe, while Planck units are more suitable for quantum gravity and early-universe cosmology. Both au and Planck units normalize the reduced Planck constant. Beyond this, Planck units normalize to 1 the two fundamental constants of general relativity and cosmology: the gravitational constant G and the speed of light in a vacuum, c. Atomic units, by contrast, normalize to 1 the mass and charge of the electron, and, as a result, the speed of light in atomic units is a large value, $1/\alpha \approx 137$. The orbital velocity of an electron around a small atom is of the order of 1 in atomic units, so the discrepancy between the velocity units in the two systems reflects the fact that electrons orbit small atoms much slower than the speed of light (around 2 orders of magnitude slower).

There are much larger discrepancies in some other units. For example, the unit of mass in atomic units is the mass of an electron, while the unit of mass in Planck units is the Planck mass, a mass so large that if a single particle had that much mass it might collapse into a black hole. Indeed, the Planck unit of mass is 22 orders of magnitude larger than the au unit of mass. Similarly, there are many orders of magnitude separating the Planck units of energy and length from the corresponding atomic units.

ATOMIC MODELS

There has been a variety of atomic models throughout history of atomic physics, that refers mainly to a period from the beginning of 19th century to the first half of 20th century, when a final model of atom which is being used nowadays (or accepted as the most accurate one) was invented. Although the awareness of atom existence goes way back to the antique period of the world history, this topic will be mainly about five basic atomic models, from which each one has somehow contributed to how we percept the structure of atom itself - Dalton´s Billiard Ball Model, J.J Thomson's "plum pudding" model, Rutherford's Planetary model, Bohr's Atomic model, Electron Cloud Model/Quantum Mechanics Model.

John Dalton's Atomic Model

John Dalton was an English scientist, who came up with an idea that all matter is composed of very small things. It was the first complete attempt to describe all matter in terms of particles. He called these particles atoms and formed an atomic theory. In this theory he claims that:

- All matter is made of atoms. Atoms are indivisible and indestructible.

- All atoms of a given element are identical in mass and properties.

- Compounds are formed by a combination of two or more different kinds of atoms.

- A chemical reaction is a rearrangement of atoms.

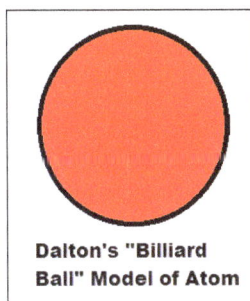

Dalton's "Billiard Ball" Model of Atom

Ilustration of Dalton's perception of atom.

Parts of his theory had to be modified based on the discovery of subatomic particles and isotopes. We now also know that atoms are not indivisible, because they are made up of neutrons, electrons and protons.

Plum Pudding Model

After discovery of an electron in 1897, people realised that atoms are made up of even smaller particles. Shortly after in 1904 J. J. Thomson proposed his famous "plum pudding model". In this model, atoms were known to consist of negatively charged electrons, however the atomic nucleus had not been discovered yet. Thomson knew that atom had an overall neutral charge. He thought that there must be something to counterbalance the negative charge of an electron. He came up with an idea that negative particles are floating within a soup of diffuse positive charge. His model is often called the plum pudding model, because of his similarity to a popular English dessert.

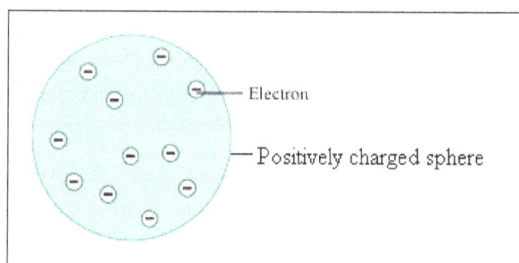

Ilustration of Thomson's perception of atom.

Rutherford's Model of the Atom

Rutherford was first, who suggested that Thomson's plum pudding model was incorrect. His new model introduces nucleus to the atom theory. Nucleus contains relatively high central charge concentrated into very small volume. This small volume also contains the bulk of the atomic mass of the atom. Nucleus is surrounded by lighter and negatively charged electrons. His model is sometimes known as the planetary model of the atom. However, there were still some major problems with this model. For example Rutherford could not explain why atoms only emit light at certain frequencies. This problem was solved later by a Danish physicist Niels Henrik David Bohr.

Bohr's Model of the Atom

Bohr model describes the atom as a positively charged nucleus, which is surrounded by electrons. Electrons travel in circular orbits, attraction is provided by electrostatic forces. Normally occupied energy level of the electron is called the ground state. The electron can move to the less – stable level by absorbing energy. This higher – energy level is called excited state. The electron can return to its original level by releasing the energy. All in all, when electron jumps between orbits, it is accompanied by an emitted or absorbed amount of energy (hv).

Electron Cloud Model/Quantum Mechanics Model of Atom

Quantum Mechanics Model of Atom is nowadays being taught as the most "realistic" atomic model that describes atomic mechanisms as how present science presumes they work. It came to exist as a result of combination of number of scientific assumptions:

• All particles could be percieved as matter waves with a wavelength.

- Resulting from the previous assumption, atomic model which treats electrons also as matter waves was proposed. (Erwin Schrödinger, quantum mechanical atomic model emerged from the solution of Schrödinger's equation for electron in central electrical field of nucleus).

- Principle of uncertainty states that we can't know both the energy and position of an electron. Therefore, as we learn more about the electron's position, we know less about its energy, and vice versa.

- There exists more than one energy level of electron in the atom. Electrons are assigned certain atomic orbitals, that can differ from one another in energy.

- Electrons have an intrinsic property called spin, and an electron can have one of two possible spin values: spin-up or spin-down. Any two electrons occupying the same orbital must have opposite spins.

Basic Description of the Quantum Mechanical Atomic Model

Quantum mechanics physics propose that electrons are moving around the nucleus not on specifically defined electron paths, but in a certain three dimensional space (atomic orbital), in which their own occurrence has a certain probability, meaning their position cannot be calculated with 100% accuracy.

Four numbers, called quantum numbers, were introduced to describe the characteristics of electrons and their orbitals.

Quantum Numbers

Principal Quantum Number: n

- Describes the energy level of the electron in an atom (it also describes the average distance of the orbital from nucleus).

- It has positive whole number values: 1, 2, 3, 4,... (theoretically speaking the numbers could go to infinite, practically there are 7 known energy levels), it can be seen sometimes described in capital letters instead of numbers, beginning with K (K, L, M, N...).

- The n value describes the size of the orbital.

Angular Momentum Quantum Number: l

- Describes basically the shape of the orbital.

- This number is limited by the principal quantum number. Its value goes from 0 to n - 1. For example, for orbitals with principal quantum number n = 2 there can by 2 different shapes of orbitals (2 different values l = 0 and l = 1).

- For every number exists a letter describing the shape of the orbital as shown in the table below.

- Value of l (subshell) Letter.

Value of l (subshell)	Letter
0	s
1	p
2	d
3	f
4	g

Magnetic Quantum Number: ml

describes how different shapes of orbitals are oriented in space. Its value can be from - l to 0 to + l. For example, for value l=1 there exists 3 values m = - 1, 0, + 1, meaning that the shape of that orbital can be oriented in 3 different ways in space.

Value of l	**Values of ml**
0	0
1	$-1, 0, +1$
3	$-1, 0, +1 - 3, -2, -1, 0, +1, +2, +3$

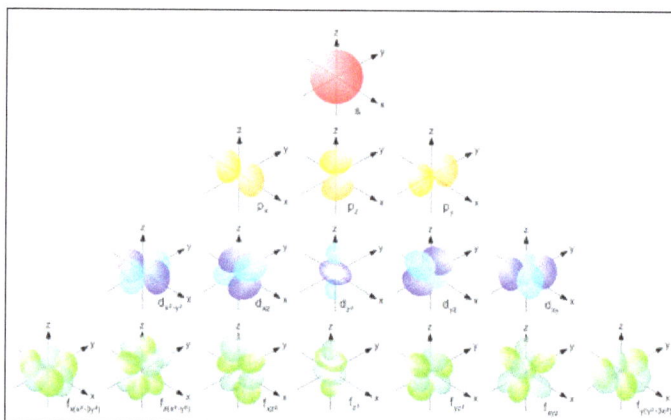

Orbital shapes and their orientation for different angular momentum and magnetic numbers.

Spin Quantum Number: ms

- Describes in which direction is an electron spinning in a magnetic field . That can be either clockwise or counterclockwise and as a result, there are only 2 values allowed: -1/2 and +1/2.

- One consequence of electron spin is that a maximum of two electrons can occupy any given orbital, and the two electrons occupying the same orbital must have opposite spin. This is also called the Pauli exclusion principle.

Principles of Atom Structure

Based on knowledge of quantum numbers, we are now able to build an electron configuration of an atom that describes electron arrangement in an atom. Apart from Pauli exclusion principle, there are 2 other rules that we must follow:

- Aufbau principle: Each electron occupies the lowest energy orbital available.

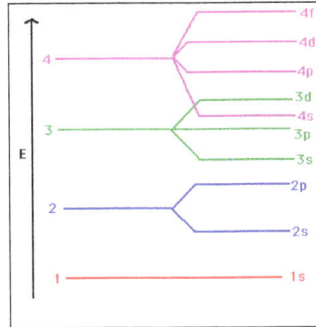

Basic ilustration of the Aufbau principle.

Hund's rule: a single electron with the same spin must occupy each orbital in a sublevel before they pair up with an electron with an opposite spin.

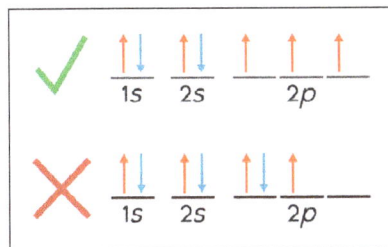

Hund's Rule of Maximum Spin Multiplicity.

	1s	2s	$2p_x$	$2p_y$	$2p_z$
H	↑				
Li	↑↓	↑			
C	↑↓	↑↓	↑	↑	
N	↑↓	↑↓	↑	↑	↑
O	↑↓	↑↓	↑↓	↑	↑

Examples of atoms described by quantum numbers.

The importance of quantum mechanical atomic model has 2 main aspects. First, being able to build an electron structure of atoms of specific substances helps us to understand how the atoms interact in molecules, therefore we are one step closer to a more detailed description of attributes of those substances. Second, it leaves an open door to more potential theories which could expand our knowledge and perception of the world and universe surrounding us.

Dalton's Atomic Model

Dalton's atomic model sets up the building blocks for others to improve on. Though some of his conclusions were incorrect, his contributions were vital. He defined an atom as the smallest indivisible particle.

Though we know today that they can be further divided into protons, neutrons, and electrons, his explanation was revolutionary for that period of time. Here's how he defined the atom:

> "Matter, though divisible in an extreme degree, is nevertheless not infinitely divisible. That is, there must be some point beyond which we cannot go in the division of matter. I have chosen the word "atom" to signify these ultimate particles."
>
> -John Dalton

John Dalton.

Basic Laws of Atomic Theory

Law of Conservation of Mass

The law of conservation of mass states that the net change in mass of the reactants and products before and after a chemical reaction is zero. This means mass can neither be created nor destroyed. In other words, the total mass in a chemical reaction remains constant.

This law was formulated by Antoine Lavoisier in 1789. It was later found to be slightly inaccurate, as in the course of chemical reactions mass can interconvert with heat and bond energy. However, these losses are very small, several orders of magnitude smaller than the mass of the reactants, so that this law is an excellent approximation.

Example:

Does the following chemical reaction obey the law of conservation of mass?

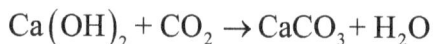

$$Ca(OH)_2 + CO_2 \rightarrow CaCO_3 + H_2O$$

The mass of Ca, O, H, and C are 40u, 16u, 1u, and 12u, respectively.

Yes, they obey the law of conservation of mass. Let's verify it. The molecular mass of

$$Ca(OH)_2 = 40 + 32 + 2$$
$$= 2$$
$$CO_2 = 12 + 32$$
$$= 44$$

$$CaCO_3 = 40+12+48$$
$$=100$$
$$H_2O = 2+16$$
$$=18$$

Substituting these values in the equation,

$$74+44=100+18$$
$$118=118.$$

Law of Constant Proportions

The law of constant proportions states that when a compound is broken, the masses of the constituent elements remain in the same proportion. Or, in a chemical compound, the elements are always present in definite proportions by mass.

It means each compound has the same elements in the same proportions, irrespective of where the compound was obtained, who prepared the compound, or the mass of the compound.

This law was formulated and proven by Joseph Louis Proust in 1799.

Example:

A person living in Australia sent a 100 ml sample of $CaCO_3$ (calcium carbonate) to a person living in India. The person living in India made his own sample of 200 ml and compared it to his friend's. Which of the two compounds has a greater ratio of Ca : C?

Both contain equal ratio of Ca and C. This is guaranteed by the law of constant proportions.

Law of Multiple Proportions

The law of multiple proportions states that when two elements form two or more compounds between them, the ratio of the masses of the second element in each compound can be expressed in the form of small whole numbers.

This law was proposed by John Dalton, and it is a combination of the previous laws.

Example

Carbon combines with oxygen to form two different compounds (under different circumstances); one is the most common gas CO_2 and the other is CO . Do they obey the law of multiple proportions?

Yes, they do obey the law of multiple proportions. Let's verify it.

We know that the mass of carbon is 12u and that of oxygen is 16 u.

So, we can say that 12 g of carbon combines with 32g of oxygen to form CO_2.

Similarly, 12g of carbon combines with 16 g of oxygen to form CO.

So, the ratio of oxygen in the first and second compound is $\dfrac{32}{16} = \dfrac{2}{1} = 2$, which is a whole number.

There is one other law which was proposed to find the relation between two different compounds.

Law of Reciprocal Proportions

The law of reciprocal proportions states that when two different elements combine with the same quantity of a third element, the ratio in which they do so will be the same or a multiple of the proportion in which they combine with each other.

This law was proposed by Jeremias Ritcher in 1792.

Dalton's Atomic Theory

Dalton picked up the idea of divisibility of matter to explain the nature of atoms. He studied the laws of chemical combinations carefully and came to a conclusion about the characteristics of atoms.

His statements were based on the three laws we'd discussed earlier. He stated the following postulates (not all of them are true) about his atomic theory.

- Matter is made of very tiny particles called atoms.

- Atoms are indivisible structures, which can neither be created nor destroyed during a chemical reaction (based on the law of conservation of mass).

- All atoms of a particular element are similar in all respects, be it their physical or chemical properties.

- Inversely, atoms of different elements show different properties, and they have different masses and different chemical properties.

- Atoms combine in the ratio of small whole numbers to form stable compounds, which is how they exist in nature.

- The relative number and the kinds of atoms in a given compound are always in a fixed ratio (based on the law of constant proportions).

As said earlier, all the postulates weren't correct. Let us discuss the drawbacks of Dalton's atomic theory.

Drawbacks

- The first part of the second postulate was not accepted. Bohr's model proposed that the atoms could be further divided into protons, neutrons, and electrons.

- The third postulate was also proven to be wrong because of the existence of isotopes, which are atoms of the same element but of different masses.

- The fourth postulate was also proven to be wrong because of the existence of isobars, which are atoms of different elements but of the same mass.

Nonetheless, to propose the idea of an atom (considering the time period) is a great achievement, and we must appreciate Dalton's work.

Dalton's Model of an Atom

Based on all his observations, Dalton proposed his model of an atom. It is often referred to as the billiard ball model. He defined an atom to be a ball-like structure, as the concepts of atomic nucleus and electrons were unknown at the time. If you asked Dalton to draw the diagram of an atom, he would've simply drawn a circle.

Later, he tried to symbolize atoms, and he became one of the first scientists to assign such symbols. He gave a specific symbol to each atom.

It was only after J. J. Thompson proposed his model that the true concepts had come into existence. Later, Rutherford worked on Dalton's and Thompson's models and brought out a roughly correct shape of the concept. Finally, Bohr's model and the quantum mechanical model gave a complete model which we know of today.

Thomson's Plum Pudding Model

Thomson atomic model was proposed by William Thomson in the year 1900. This model explained the description of an inner structure of the atom theoretically. It was strongly supported by Sir Joseph Thomson, who had discovered the electron earlier.

During cathode ray tube experiment, a negatively charged particle was discovered by J.J. Thomson. This experiment took place in the year 1897. Cathode ray tube is a vacuum tube. The negative particle was called an electron.

Thomson assumed that an electron is two thousand times lighter than a proton and believed that an atom is made up of thousands of electrons. In this atomic structure model, he considered atoms surrounded by a cloud having positive as well as negative charges. The demonstration of the ionization of air by X-ray was also done by him together with Rutherford. They were the first to demonstrate it. Thomson's model of an atom is similar to a plum pudding.

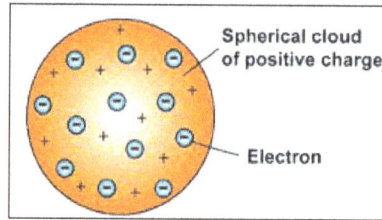

Postulates of Thomson's Atomic Model

Postulate 1: An atom consists of a positively charged sphere with electrons embedded in it.

Postulate 2: An atom as a whole is electrically neutral because the negative and positive charges are equal in magnitude.

Thomson atomic model is compared to watermelon. Where he considered:

- Watermelon seeds as negatively charged particles.

- The red part of the watermelon as positively charged.

Limitations of Thomson's Atomic Model

- It failed to explain the stability of an atom because his model of atom failed to explain how a positive charge holds the negatively charged electrons in an atom. Therefore, This theory also failed to account for the position of the nucleus in an atom.

- Thomson's model failed to explain the scattering of alpha particles by thin metal foils.

- No experimental evidence in its support.

Although Thomson's model was not an accurate model to account for the atomic structure, it proved to be the base for the development of other atomic models.

Rutherford Model

Rutherford proposed that an atom is composed of empty space mostly with electrons orbiting in a set, predictable paths around fixed, positively charged nucleus.

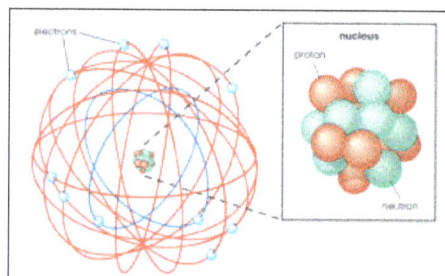

Rutherford's Atomic Model.

The concept of atom dates back to 400 BCE when Greek philosopher Democritus first conceived the idea. However, it was not until 1803 John Dalton proposed again the idea of the atom. But at that point of time, atoms were considered indivisible. This idea of an atom as indivisible particles continued until the year 1897 when British Physicist J.J. Thomson discovered negatively charged particles which were later named electrons.

He proposed a model on the basis of that where he explained electrons were embedded uniformly in a positively charged matrix. The model was named plum pudding model. However, J.J. Thomson's plum pudding model had some limitations. It failed to explain certain experimental results related to the atomic structure of elements.

A British Physicist "Ernest Rutherford" proposed a model of the atomic structure known as Rutherford's Model of Atoms. He conducted an experiment where he bombarded α-particles in a thin sheet of gold. In this experiment, he studied the trajectory of the α-particles after interaction with the thin sheet of gold.

Rutherford Atomic Model Experiment

In Rutherford's experiment, he bombarded high energy streams of α-particles on a thin gold foil of 100 nm thickness. The streams of α-particles were directed from a radioactive source. He conducted the experiment to study the deflection produced in the trajectory of α-particles after interaction with the thin sheet of gold. To study the deflection, he placed a screen made up of zinc sulfide around the gold foil. The observations made by Rutherford contradicted the plum pudding model given by J.J. Thomson.

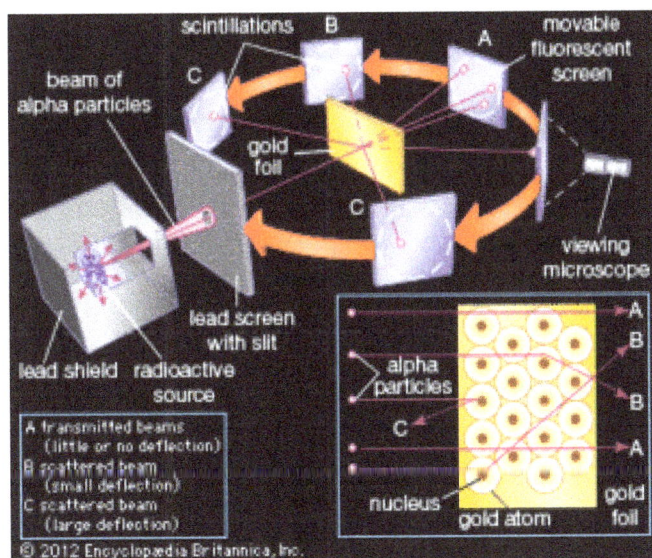

Rutherford's Gold Foil Experiment.

Observations of Rutherford Model Experiment

On the basis of the observations made during the experiment, Rutherford concluded that:

- Major space in an atom is empty – A large fraction of α-particles passed through the gold sheet without getting deflected. Therefore, the major part of an atom must be empty.

- The positive charge in an atom is not distributed uniformly and it is concentrated in a very small volume – Few α-particles when bombarded were deflected by the gold sheet. They were deflected minutely and at very small angles. Therefore he made the above conclusion.

- Very few α-particles had deflected at large angles or deflected back. Moreover, very few particles had deflected at 180o. Therefore, he concluded that the positively charged particles covered a small volume of an atom in comparison to the total volume of an atom.

Postulates of Rutherford Atomic Model based on Observations and Conclusions

- An atom is composed of positively charged particles. Majority of the mass of an atom was concentrated in a very small region. This region of the atom was called as the nucleus of an atom. It was found out later that the very small and dense nucleus of an atom is composed of neutrons and protons.

- Atoms nucleus is surrounded by negatively charged particles called electrons. The electrons revolve around the nucleus in a fixed circular path at very high speed. These fixed circular paths were termed as "orbits."

- An atom has no net charge or they are electrically neutral because electrons are negatively charged and the densely concentrated nucleus is positively charged. A strong electrostatic force of attractions holds together the nucleus and electrons.

- The size of the nucleus of an atom is very small in comparison to the total size of an atom.

Limitations of Rutherford Atomic Model

Rutherford's experiment was unable to explain certain things. They are:

- Rutherford's model was unable to explain the stability of an atom. According to Rutherford's postulate, electrons revolve at a very high speed around a nucleus of an atom in a fixed orbit. However, Maxwell explained accelerated charged particles release electromagnetic radiations. Therefore, electrons revolving around the nucleus will release electromagnetic radiation.

- The electromagnetic radiation will have energy from the electronic motion as a result of which the orbits will gradually shrink. Finally, the orbits will shrink and collapse in the nucleus of an atom. According to the calculations, if Maxwell's explanation is followed Rutherford's model will collapse with 10-8 seconds. Therefore, Rutherford atomic model was not following Maxwell's theory and it was unable to explain an atom's stability.

- Rutherford's theory was incomplete because it did not mention anything about the arrangement of electrons in the orbit. This was one of the major drawbacks of Rutherford atomic model.

Even though the early atomic models were inaccurate and could not explain the structure of atom and experimental results properly. But it formed the basis of the quantum mechanics and helped the future development of quantum mechanics.

NUCLEAR BINDING ENERGY

Nuclear binding energy is the minimum energy that would be required to disassemble the nucleus of an atom into its component parts. These component parts are neutrons and protons, which are collectively called nucleons. The binding is always a positive number, as we need to spend energy in moving these nucleons, attracted to each other by the strong nuclear force, away from each other. The mass of an atomic nucleus is less than the sum of the individual masses of the free constituent protons and neutrons, according to Einstein's equation E=mc². This 'missing mass' is known as the mass defect, and represents the energy that was released when the nucleus was formed.

The term "nuclear binding energy" may also refer to the energy balance in processes in which the nucleus splits into fragments composed of more than one nucleon. If new binding energy is available when light nuclei fuse (nuclear fusion), or when heavy nuclei split (nuclear fission), either process can result in release of this binding energy. This energy may be made available as nuclear energy and can be used to produce electricity, as in nuclear power, or in a nuclear weapon. When a large nucleus splits into pieces, excess energy is emitted as photon (gamma rays) and as the kinetic energy of a number of different ejected particles (nuclear fission products).

These nuclear binding energies and forces are on the order of a million times greater than the electron binding energies of light atoms like hydrogen.

The mass defect of a nucleus represents the amount of mass equivalent to the binding energy of the nucleus (E=mc²), which is the difference between the mass of a nucleus and the sum of the individual masses of the nucleons of which it is composed.

Nuclear binding energy is explained by the basic principles involved in nuclear physics.

Nuclear Energy

An absorption or release of nuclear energy occurs in nuclear reactions or radioactive decay; those that absorb energy are called endothermic reactions and those that release energy are exothermic reactions. Energy is consumed or liberated because of differences in the nuclear binding energy between the incoming and outgoing products of the nuclear transmutation.

The best-known classes of exothermic nuclear transmutations are fission and fusion. Nuclear energy may be liberated by atomic fission, when heavy atomic nuclei (like uranium and plutonium) are broken apart into lighter nuclei. The energy from fission is used to generate electric power in hundreds of locations worldwide. Nuclear energy is also released during atomic fusion, when light nuclei like hydrogen are combined to form heavier nuclei such as helium. The Sun and other stars use nuclear fusion to generate thermal energy which is later radiated from the surface, a type of stellar nucleosynthesis. In any exothermic nuclear process, nuclear mass might ultimately be converted to thermal energy, given off as heat.

In order to quantify the energy released or absorbed in any nuclear transmutation, one must know the nuclear binding energies of the nuclear components involved in the transmutation.

The Nuclear Force

Electrons and nuclei are kept together by electrostatic attraction (negative attracts positive). Furthermore, electrons are sometimes shared by neighboring atoms or transferred to them (by processes of quantum physics), and this link between atoms is referred to as a chemical bond, and is responsible for the formation of all chemical compounds.

The force of electric attraction does not hold nuclei together, because all protons carry a positive charge and repel each other. Thus, electric forces do not hold nuclei together, because they act in the opposite direction. It has been established that binding neutrons to nuclei clearly requires a non-electrical attraction.

Therefore, another force, called the nuclear force (or residual strong force) holds the nucleons of nuclei together. This force is a residuum of the strong interaction, which binds quarks into nucleons at an even smaller level of distance.

The nuclear force must be stronger than the electric repulsion at short distances, but weaker far away, or else different nuclei might tend to clump together. Therefore, it has short-range characteristics. An analogy to the nuclear force is the force between two small magnets: magnets are very difficult to separate when stuck together, but once pulled a short distance apart, the force between them drops almost to zero.

Unlike gravity or electrical forces, the nuclear force is effective only at very short distances. At greater distances, the electrostatic force dominates: the protons repel each other because they are positively charged, and like charges repel. For that reason, the protons forming the nuclei of ordinary hydrogen—for instance, in a balloon filled with hydrogen—do not combine to form helium (a process that also would require some protons to combine with electrons and become neutrons). They cannot get close enough for the nuclear force, which attracts them to each other, to become important. Only under conditions of extreme pressure and temperature (for example, within the core of a star), can such a process take place.

Physics of Nuclei

There are around 94 naturally occurring elements on earth. The atoms of each element have a nucleus containing a specific number of protons (always the same number for a given element), and some number of neutrons, which is often roughly a similar number. Two atoms of the same element having different numbers of neutrons are known as isotopes of the element. Different isotopes may have different properties - for example one might be stable and another might be unstable, and gradually undergo radioactive decay to become another element.

The hydrogen nucleus contains just one proton. Its isotope deuterium, or heavy hydrogen, contains a proton and a neutron. Helium contains two protons and two neutrons, and carbon, nitrogen and oxygen - six, seven and eight of each particle, respectively. However, a helium nucleus weighs less than the sum of the weights of the two heavy hydrogen nuclei which combine to make it. The same is true for carbon, nitrogen and oxygen. For example, the carbon nucleus is slightly lighter than three helium nuclei, which can combine to make a carbon nucleus. This difference is known as the mass defect.

Mass Defect

Mass defect is the difference between the mass of an object and the sum of the masses of its constituent particles. Discovered by Albert Einstein in 1905, it can be explained using his formula E = mc², which describes the equivalence of energy and mass. The decrease in mass is equal to the energy given off in the reaction of an atom's creation divided by c². By this formula, adding energy also increases mass (both weight and inertia), whereas removing energy decreases mass. For example, a helium atom containing four nucleons has a mass about 0.8% less than the total mass of four hydrogen nuclei (which contain one nucleon each). The helium nucleus has four nucleons bound together, and the binding energy which holds them together is, in effect, the missing 0.8% of mass.

If a combination of particles contains extra energy—for instance, in a molecule of the explosive TNT—weighing it reveals some extra mass, compared to its end products after an explosion. (The weighing must be done after the products have been stopped and cooled, however, as the extra mass must escape from the system as heat before its loss can be noticed, in theory.) On the other hand, if one must inject energy to separate a system of particles into its components, then the initial mass is less than that of the components after they are separated. In the latter case, the energy injected is "stored" as potential energy, which shows as the increased mass of the components that store it. This is an example of the fact that energy of all types is seen in systems as mass, since mass and energy are equivalent, and each is a "property" of the other.

The latter scenario is the case with nuclei such as helium: to break them up into protons and neutrons, one must inject energy. On the other hand, if a process existed going in the opposite direction, by which hydrogen atoms could be combined to form helium, then energy would be released. The energy can be computed using $E = \Delta m c^2$ for each nucleus, where Δm is the difference between the mass of the helium nucleus and the mass of four protons (plus two electrons, absorbed to create the neutrons of helium).

For lighter elements, the energy that can be released by assembling them from lighter elements decreases, and energy can be released when they fuse. This is true for nuclei lighter than iron/nickel. For heavier nuclei, more energy is needed to bind them, to the point that energy is released by breaking them up into 2 fragments (known as atomic fission). Nuclear power is generated at present by breaking up uranium nuclei in nuclear power reactors, and capturing the released energy as heat, which is converted to electricity.

As a rule, very light elements can fuse comparatively easily, and very heavy elements can break up via fission very easily; elements in the middle are more stable and it is difficult to make them undergo either fusion or fission in an earthly environment such as a laboratory.

The reason the trend reverses after iron is the growing positive charge of the nuclei, which tends to force nuclei to break up. It is resisted by the strong nuclear interaction, which holds nucleons together. The electric force may be weaker than the strong nuclear force, but the strong force has a much more limited range: in an iron nucleus, each proton repels the other 25 protons, while the nuclear force only binds close neighbors. So for larger nuclei, the electrostatic forces tend to dominate and the nucleus will tend over time to break up.

As nuclei grow bigger still, this disruptive effect becomes steadily more significant. By the time polonium is reached (84 protons), nuclei can no longer accommodate their large positive charge, but emit their excess protons quite rapidly in the process of alpha radioactivity—the emission of helium nuclei, each containing two protons and two neutrons. (Helium nuclei are an especially stable combination.) Because of this process, nuclei with more than 94 protons are not found naturally on Earth. The isotopes beyond uranium (atomic number 92) with the longest half-lives are plutonium-244 (80 million years) and curium-247 (16 million years).

Solar Binding Energy

The nuclear fusion process works as follows: five billion years ago, the new Sun formed when gravity pulled together a vast cloud of hydrogen and dust, from which the Earth and other planets also arose. The gravitational pull released energy and heated the early Sun, much in the way Helmholtz proposed.

Thermal energy appears as the motion of atoms and molecules: the higher the temperature of a collection of particles, the greater is their velocity and the more violent are their collisions. When the temperature at the center of the newly formed Sun became great enough for collisions between hydrogen nuclei to overcome their electric repulsion, and bring them into the short range of the attractive nuclear force, nuclei began to stick together. When this began to happen, protons combined into deuterium and then helium, with some protons changing in the process to neutrons (plus positrons, positive electrons, which combine with electrons and annihilate into gamma-ray photons). This released nuclear energy now keeps up the high temperature of the Sun's core, and the heat also keeps the gas pressure high, keeping the Sun at its present size, and stopping gravity from compressing it any more. There is now a stable balance between gravity and pressure.

Different nuclear reactions may predominate at different stages of the Sun's existence, including the proton-proton reaction and the carbon-nitrogen cycle—which involves heavier nuclei, but whose final product is still the combination of protons to form helium.

A branch of physics, the study of controlled nuclear fusion, has tried since the 1950s to derive useful power from nuclear fusion reactions that combine small nuclei into bigger ones, typically to heat boilers, whose steam could turn turbines and produce electricity. Unfortunately, no earthly laboratory can match one feature of the solar powerhouse: the great mass of the Sun, whose weight keeps the hot plasma compressed and confines the nuclear furnace to the Sun's core. Instead, physicists use strong magnetic fields to confine the plasma, and for fuel they use heavy forms of hydrogen, which burn more easily. Magnetic traps can be rather unstable, and any plasma hot enough and dense enough to undergo nuclear fusion tends to slip out of them after a short time. Even with ingenious tricks, the confinement in most cases lasts only a small fraction of a second.

Combining Nuclei

Small nuclei that are larger than hydrogen can combine into bigger ones and release energy, but in combining such nuclei, the amount of energy released is much smaller compared to hydrogen fusion. The reason is that while the overall process releases energy from letting the nuclear attraction do its work, energy must first be injected to force together positively charged protons, which also repel each other with their electric charge.

For elements that weigh more than iron (a nucleus with 26 protons), the fusion process no longer releases energy. In even heavier nuclei energy is consumed, not released, by combining similarly sized nuclei. With such large nuclei, overcoming the electric repulsion (which affects all protons in the nucleus) requires more energy than is released by the nuclear attraction (which is effective mainly between close neighbors). Conversely, energy could actually be released by breaking apart nuclei heavier than iron.

With the nuclei of elements heavier than lead, the electric repulsion is so strong that some of them spontaneously eject positive fragments, usually nuclei of helium that form very stable combinations (alpha particles). This spontaneous break-up is one of the forms of radioactivity exhibited by some nuclei.

Nuclei heavier than lead (except for bismuth, thorium, and uranium) spontaneously break up too quickly to appear in nature as primordial elements, though they can be produced artificially or as intermediates in the decay chains of heavier elements. Generally, the heavier the nuclei are, the faster they spontaneously decay.

Iron nuclei are the most stable nuclei (in particular iron-56), and the best sources of energy are therefore nuclei whose weights are as far removed from iron as possible. One can combine the lightest ones—nuclei of hydrogen (protons)—to form nuclei of helium, and that is how the Sun generates its energy. Or else one can break up the heaviest ones—nuclei of uranium or plutonium—into smaller fragments, and that is what nuclear reactors do.

Nuclear Binding Energy

An example that illustrates nuclear binding energy is the nucleus of ^{12}C (carbon-12), which contains 6 protons and 6 neutrons. The protons are all positively charged and repel each other, but the nuclear force overcomes the repulsion and causes them to stick together. The nuclear force is a close-range force (it is strongly attractive at a distance of 1.0 fm and becomes extremely small beyond a distance of 2.5fm), and virtually no effect of this force is observed outside the nucleus. The nuclear force also pulls neutrons together, or neutrons and protons.

The energy of the nucleus is negative with regard to the energy of the particles pulled apart to infinite distance (just like the gravitational energy of planets of the solar system), because energy must be utilized to split a nucleus into its individual protons and neutrons. Mass spectrometers have measured the masses of nuclei, which are always less than the sum of the masses of protons and neutrons that form them, and the difference—by the formula $E = mc^2$ gives the binding energy of the nucleus.

Nuclear Fusion

The binding energy of helium is the energy source of the Sun and of most stars. The sun is composed of 74 percent hydrogen (measured by mass), an element having a nucleus consisting of a single proton. Energy is released in the sun when 4 protons combine into a helium nucleus, a process in which two of them are also converted to neutrons.

The conversion of protons to neutrons is the result of another nuclear force, known as the weak (nuclear) force. The weak force, like the strong force, has a short range, but is much weaker than the strong force. The weak force tries to make the number of neutrons and protons into the most energetically stable configuration. For nuclei containing less than 40 particles, these numbers are usually about equal. Protons and neutrons are closely related and are collectively known as nucleons. As the number of particles increases toward a maximum of about 209, the number of neutrons to maintain stability begins to outstrip the number of protons, until the ratio of neutrons to protons is about three to two.

The protons of hydrogen combine to helium only if they have enough velocity to overcome each other's mutual repulsion sufficiently to get within range of the strong nuclear attraction. This means that fusion only occurs within a very hot gas. Hydrogen hot enough for combining to helium requires an enormous pressure to keep it confined, but suitable conditions exist in the central regions of the Sun, where such pressure is provided by the enormous weight of the layers above the core, pressed inwards by the Sun's strong gravity. The process of combining protons to form helium is an example of nuclear fusion.

The earth's oceans contain a large amount of hydrogen that could theoretically be used for fusion, and helium byproduct of fusion does not harm the environment, so some consider nuclear fusion a good alternative to supply humanity's energy needs. Experiments to generate electricity from fusion have so far only partially succeeded. Sufficiently hot hydrogen must be ionized and confined. One technique is to use very strong magnetic fields, because charged particles (like those trapped in the Earth's radiation belt) are guided by magnetic field lines. Fusion experiments also rely on heavy hydrogen, which fuses more easily, and gas densities can be moderate. But even with these techniques far more net energy is consumed by the fusion experiments than is yielded by the process.

The Binding Energy Maximum and Ways to Approach it by Decay

In the main isotopes of light nuclei, such as carbon, nitrogen and oxygen, the most stable combination of neutrons and of protons are when the numbers are equal (this continues to element 20, calcium). However, in heavier nuclei, the disruptive energy of protons increases, since they are confined to a tiny volume and repel each other. The energy of the strong force holding the nucleus together also increases, but at a slower rate, as if inside the nucleus, only nucleons close to each other are tightly bound, not ones more widely separated.

The net binding energy of a nucleus is that of the nuclear attraction, minus the disruptive energy of the electric force. As nuclei get heavier than helium, their net binding energy per nucleon (deduced from the difference in mass between the nucleus and the sum of masses of component nucleons) grows more and more slowly, reaching its peak at iron. As nucleons are added, the total nuclear binding energy always increases—but the total disruptive energy of electric forces (positive protons repelling other protons) also increases, and past iron, the second increase outweighs the first. Iron-56 (^{56}Fe) is the most efficiently bound nucleus meaning that it has the least average mass per nucleon. However, nickel-62 is the most tightly bound nucleus in terms of energy of binding per nucleon . (Nickel-62's higher energy of binding does not translate to a larger mean mass loss than Fe-56, because Ni-62 has a slightly higher ratio of neutrons/protons than does iron-56, and the presence of the heavier neutrons increases nickel-62's average mass per nucleon).

To reduce the disruptive energy, the weak interaction allows the number of neutrons to exceed that of protons—for instance, the main isotope of iron has 26 protons and 30 neutrons. Isotopes also exist where the number of neutrons differs from the most stable number for that number of nucleons. If the ratio of protons to neutrons is too far from stability, nucleons may spontaneously change from proton to neutron, or neutron to proton.

The two methods for this conversion are mediated by the weak force, and involve types of beta decay. In the simplest beta decay, neutrons are converted to protons by emitting a negative electron and an antineutrino. This is always possible outside a nucleus because neutrons are more massive than protons by an equivalent of about 2.5 electrons. In the opposite process, which only happens within a nucleus, and not to free particles, a proton may become a neutron by ejecting a positron. This is permitted if enough energy is available between parent and daughter nuclides to do this (the required energy difference is equal to 1.022 MeV, which is the mass of 2 electrons). If the mass difference between parent and daughter is less than this, a proton-rich nucleus may still convert protons to neutrons by the process of electron capture, in which a proton simply electron captures one of the atom's K orbital electrons, emits a neutrino, and becomes a neutron.

Among the heaviest nuclei, starting with tellurium nuclei (element 52) containing 106 or more nucleons, electric forces may be so destabilizing that entire chunks of the nucleus may be ejected, usually as alpha particles, which consist of two protons and two neutrons (alpha particles are fast helium nuclei). (Beryllium-8 also decays, very quickly, into two alpha particles.) Alpha particles are extremely stable. This type of decay becomes more and more probable as elements rise in atomic weight past 106.

The curve of binding energy is a graph that plots the binding energy per nucleon against atomic mass. This curve has its main peak at iron and nickel and then slowly decreases again, and also a narrow isolated peak at helium, which as noted is very stable. The heaviest nuclei in nature, uranium ^{238}U, are unstable, but having a half-life of 4.5 billion years, close to the age of the Earth, they are still relatively abundant; they (and other nuclei heavier than helium) have formed in stellar evolution events like supernova explosions preceding the formation of the solar system. The most common isotope of thorium, ^{232}Th, also undergoes alpha particle emission, and its half-life (time over which half a number of atoms decays) is even longer, by several times. In each of these, radioactive decay produces daughter isotopes that are also unstable, starting a chain of decays that ends in some stable isotope of lead.

Determining Nuclear Binding Energy

Calculation can be employed to determine the nuclear binding energy of nuclei. The calculation involves determining the mass defect, converting it into energy, and expressing the result as energy per mole of atoms, or as energy per nucleon.

Conversion of Mass Defect into Energy

Mass defect is defined as the difference between the mass of a nucleus, and the sum of the masses of the nucleons of which it is composed. The mass defect is determined by calculating three quantities. These are: the actual mass of the nucleus, the composition of the nucleus (number of protons and of neutrons), and the masses of a proton and of a neutron. This is then followed by converting the mass defect into energy. This quantity is the nuclear binding energy, however it must be expressed as energy per mole of atoms or as energy per nucleon.

Fission and Fusion

Nuclear energy is released by the splitting (fission) or merging (fusion) of the nuclei of atom(s). The conversion of nuclear mass-energy to a form of energy, which can remove some mass when the energy is removed, is consistent with the mass-energy equivalence formula:

$$\Delta E = \Delta m\, c^2,$$

in which,

ΔE = energy release,

Δm = mass defect,

and c = the speed of light in a vacuum (a physical constant 299,792,458 m/s by definition).

Nuclear energy was first discovered by French physicist Henri Becquerel in 1896, when he found that photographic plates stored in the dark near uranium were blackened like X-ray plates (X-rays had recently been discovered in 1895).

Nickel-62 has the highest binding energy per nucleon of any isotope. If an atom of lower average binding energy is changed into two atoms of higher average binding energy, energy is given off. Also, if two atoms of lower average binding energy fuse into an atom of higher average binding energy, energy is given off. The chart shows that fusion of hydrogen, the combination to form heavier atoms, releases energy, as does fission of uranium, the breaking up of a larger nucleus into smaller parts. Stability varies between isotopes: the isotope U-235 is much less stable than the more common U-238.

Nuclear energy is released by three exoenergetic (or exothermic) processes:

- Radioactive decay, where a neutron or proton in the radioactive nucleus decays spontaneously by emitting either particles, electromagnetic radiation (gamma rays), or both. Note that for radioactive decay, it is not strictly necessary for the binding energy to increase. What is strictly necessary is that the mass decrease. If a neutron turns into a proton and the energy of the decay is less than 0.782343 MeV (such as rubidium-87 decaying to strontium-87), the average binding energy per nucleon will actually decrease.

- Fusion, two atomic nuclei fuse together to form a heavier nucleus.

- Fission, the breaking of a heavy nucleus into two (or more rarely three) lighter nuclei.

Binding Energy for Atoms

The binding energy of an atom (including its electrons) is not the same as the binding energy of the atom's nucleus. The measured mass deficits of isotopes are always listed as mass deficits of the neutral atoms of that isotope, and mostly in MeV. As a consequence, the listed mass deficits are not a measure for the stability or binding energy of isolated nuclei, but for the whole atoms. This has very practical reasons, because it is very hard to totally ionize heavy elements, i.e. strip them of all of their electrons.

This practice is useful for other reasons, too: stripping all the electrons from a heavy unstable nucleus (thus producing a bare nucleus) changes the lifetime of the nucleus, or the nucleus of a stable neutral atom can likewise become unstable after stripping, indicating that the nucleus cannot be treated independently. Examples of this have been shown in bound-state β decay experiments performed at the GSI) heavy ion accelerator. This is also evident from phenomena like electron capture. Theoretically, in orbital models of heavy atoms, the electron orbits partially inside the nucleus (it does not orbit in a strict sense, but has a non-vanishing probability of being located inside the nucleus).

A nuclear decay happens to the nucleus, meaning that properties ascribed to the nucleus change in the event. In the field of physics the concept of "mass deficit" as a measure for "binding energy" means "mass deficit of the neutral atom" (not just the nucleus) and is a measure for stability of the whole atom.

Nuclear Binding Energy Curve

In the periodic table of elements, the series of light elements from hydrogen up to sodium is observed to exhibit generally increasing binding energy per nucleon as the atomic mass increases. This increase is generated by increasing forces per nucleon in the nucleus, as each additional nucleon is attracted by other nearby nucleons, and thus more tightly bound to the whole.

The region of increasing binding energy is followed by a region of relative stability (saturation) in the sequence from magnesium through xenon. In this region, the nucleus has become large enough that nuclear forces no longer completely extend efficiently across its width. Attractive nuclear forces in this region, as atomic mass increases, are nearly balanced by repellent electromagnetic forces between protons, as the atomic number increases.

Finally, in elements heavier than xenon, there is a decrease in binding energy per nucleon as atomic number increases. In this region of nuclear size, electromagnetic repulsive forces are beginning to overcome the strong nuclear force attraction.

At the peak of binding energy, nickel-62 is the most tightly bound nucleus (per nucleon), followed by iron-58 and iron-56. This is the approximate basic reason why iron and nickel are very common metals in planetary cores, since they are produced profusely as end products in supernovae and in the final stages of silicon burning in stars. However, it is not binding energy per defined nucleon (as defined above), which controls which exact nuclei are made, because within stars, neutrons are free to convert to protons to release even more energy, per generic nucleon, if the result is a stable nucleus with a larger fraction of protons. In fact, it has been argued that photodisintegration of ^{62}Ni to form ^{56}Fe may be energetically possible in an extremely hot star core, due to this beta decay conversion of neutrons to protons. The conclusion is that at the pressure and temperature conditions in the cores of large stars, energy is released by converting all matter into ^{56}Fe nuclei (ionized atoms). (However, at high temperatures not all matter will be in the lowest energy state.) This energetic maximum should also hold for ambient conditions, say $T = 298$ K and $p = 1$ atm, for neutral condensed matter consisting of ^{56}Fe atoms—however, in these conditions nuclei of atoms are inhibited from fusing into the most stable and low energy state of matter.

It is generally believed that iron-56 is more common than nickel isotopes in the universe for mechanistic reasons, because its unstable progenitor nickel-56 is copiously made by staged build-up of 14 helium nuclei inside supernovas, where it has no time to decay to iron before being released into the interstellar medium in a matter of a few minutes, as the supernova explodes. However, nickel-56 then decays to cobalt-56 within a few weeks, then this radioisotope finally decays to iron-56 with a half life of about 77.3 days. The radioactive decay-powered light curve of such a process has been observed to happen in type II supernovae, such as SN 1987A. In a star, there are no good ways to create nickel-62 by alpha-addition processes, or else there would presumably be more of this highly stable nuclide in the universe.

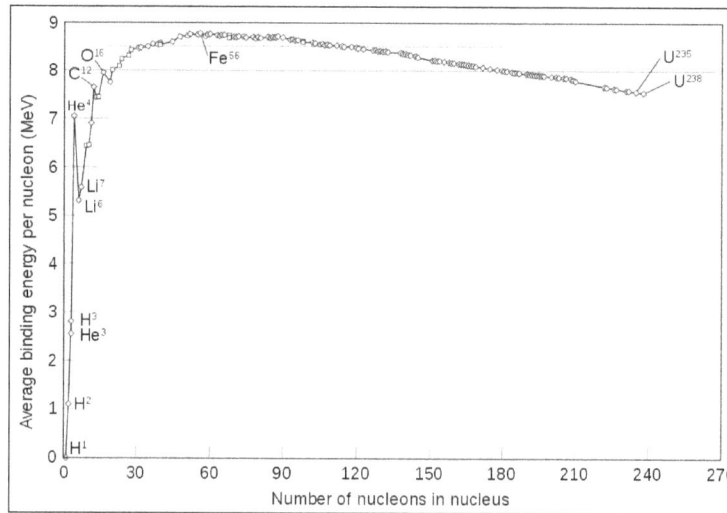

Binding Energy and Nuclide Masses

The fact that the maximum binding energy is found in medium-sized nuclei is a consequence of the trade-off in the effects of two opposing forces that have different range characteristics. The attractive nuclear force (strong nuclear force), which binds protons and neutrons equally to each other, has a limited range due to a rapid exponential decrease in this force with distance. However, the repelling electromagnetic force, which acts between protons to force nuclei apart, falls off with distance much more slowly (as the inverse square of distance). For nuclei larger than about four nucleons in diameter, the additional repelling force of additional protons more than offsets any binding energy that results between further added nucleons as a result of additional strong force interactions. Such nuclei become increasingly less tightly bound as their size increases, though most of them are still stable. Finally, nuclei containing more than 209 nucleons (larger than about 6 nucleons in diameter) are all too large to be stable, and are subject to spontaneous decay to smaller nuclei.

Nuclear fusion produces energy by combining the very lightest elements into more tightly bound elements (such as hydrogen into helium), and nuclear fission produces energy by splitting the heaviest elements (such as uranium and plutonium) into more tightly bound elements (such as barium and krypton). Both processes produce energy, because middle-sized nuclei are the most tightly bound of all.

As seen in the example of deuterium, nuclear binding energies are large enough that they may be easily measured as fractional mass deficits, according to the equivalence of mass and energy. The atomic binding energy is simply the amount of energy (and mass) released, when a collection of free nucleons are joined together to form a nucleus.

Nuclear binding energy can be computed from the difference in mass of a nucleus, and the sum of the masses of the number of free neutrons and protons that make up the nucleus. Once this mass difference, called the mass defect or mass deficiency, is known, Einstein's mass-energy equivalence formula E = mc² can be used to compute the binding energy of any nucleus. Early nuclear physicists used to refer to computing this value as a "packing fraction" calculation.

For example, the atomic mass unit (1 u) is defined as 1/12 of the mass of a ¹²C atom—but the atomic mass of a ¹H atom (which is a proton plus electron) is 1.007825 u, so each nucleon in ¹²C has lost, on average, about 0.8% of its mass in the form of binding energy.

For a nucleus with A nucleons, including Z protons and N neutrons, a semi-empirical formula for the binding energy (BE) per nucleon is:

$$\frac{BE}{A \cdot MeV} = a - \frac{b}{A^{1/3}} - \frac{cZ^2}{A^{4/3}} - \frac{d(N-Z)^2}{A^2} \pm \frac{e}{A^{7/4}}$$

where the coefficients are given by: a = 14.0; b = 13.0; c = 0.585; d = 19.3; e = 33.

The first term is called the saturation contribution and ensures that the binding energy per nucleon is the same for all nuclei to a first approximation. The term $-b/A^{1/3}$ is a surface tension effect and is proportional to the number of nucleons that are situated on the nuclear surface; it is largest for light nuclei. The term $cZ^2 \; A^{4/3}$ is the Coulomb electrostatic repulsion; this becomes more important as Zincreases. The symmetry correction term $-d(N-Z)^2/A^2$ takes into account the fact that in the absence of other effects the most stable arrangement has equal numbers of protons and neutrons; this is because the n-p interaction in a nucleus is stronger than either the n-n or p-p interaction. The pairing term $\pm e/A^{7/4}$ is purely empirical; it is + for even-even nuclei and - for odd-odd nuclei.

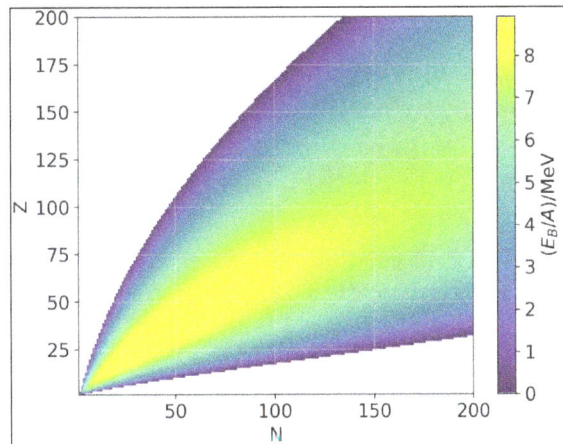

A graphical representation of the semi-empirical binding energy formula. The binding energy per nucleon in MeV (highest numbers in yellow, in excess of 8.5 MeV per nucleon) is plotted for various nuclides as a function of Z, the atomic number (y-axis), vs. N, the number of neutrons (x-axis). The highest numbers are seen for Z = 26 (iron).

Example Values Deduced from Experimentally Measured Atom Nuclide Masses

The following table lists some binding energies and mass defect values. Notice also that we use 1 u = (931.494028 ± 0.000023) MeV. To calculate the binding energy we use the formula $Z(m_p + m_e) + N m_n - m_{nuclide}$ where Z denotes the number of protons in the nuclides and N their number of neutrons. We take m_p = (938.2720813±0.0000058) MeV, m_e = (0.5109989461±0.000000003) MeV and m_n = (939.5654133 ± 0000058) MeV. The letter A denotes the sum of Z and N (number of nucleons in the nuclide). If we assume the reference nucleon has the mass of a neutron (so that all "total" binding energies calculated are maximal) we could define the total binding energy as the difference from the mass of the nucleus, and the mass of a collection of A free neutrons. In other words, it would be $(Z+N)m_n - m_{nuclide}$. The "total binding energy per nucleon" would be this value divided by A.

Table: Most strongly bound nuclides atoms.

nu-clide	Z	N	mass excess	total mass	total mass / A	total binding energy / A	mass defect	binding energy	binding energy / A
^{56}Fe	26	30	−60.6054 MeV	55.934937 u	0.9988372 u	9.1538 MeV	0.528479 u	492.275 MeV	8.7906 MeV
^{56}Fe	26	32	−62.1534 MeV	57.932276 u	0.9988496 u	9.1432 MeV	0.547471 u	509.966 MeV	8.7925 MeV
^{60}Ni	28	32	−64.472 MeV	59.93079 u	0.9988464 u	9.1462 MeV	0.565612 u	526.864 MeV	8.7811 MeV
^{62}Ni	28	34	−66.7461 MeV	61.928345 u	0.9988443 u	9.1481 MeV	0.585383 u	545.281 MeV	8.7948 MeV

^{56}Fe has the lowest nucleon-specific mass of the four nuclides listed in this table, but this does not imply it is the strongest bound atom per hadron, unless the choice of beginning hadrons is completely free. Iron releases the largest energy if any 56 nucleons are allowed to build a nuclide—changing one to another if necessary, The highest binding energy per hadron, with the hadrons starting as the same number of protons Z and total nucleons A as in the bound nucleus, is ^{62}Ni. Thus, the true absolute value of the total binding energy of a nucleus depends on what we are allowed to construct the nucleus out of. If all nuclei of mass number A were to be allowed to be constructed of A neutrons, then ^{56}Fe would release the most energy per nucleon, since it has a larger fraction of protons than 62Ni. However, if nuclei are required to be constructed of only the same number of protons and neutrons that they contain, then nickel-62 is the most tightly bound nucleus, per nucleon.

Table: Some light nuclides resp. atoms.

nuclide	Z	N	mass excess	total mass	total mass / A	total binding energy / A	mass defect	binding energy	binding energy / A
n	0	1	8.0716 MeV	1.008665 u	1.008665 u	0.0000 MeV	0 u	0 MeV	0 MeV
^{1}H	1	0	7.2890 MeV	1.007825 u	1.007825 u	0.7826 MeV	0.0000000146 u	0.0000136 MeV	13.6 eV
^{2}H	1	1	13.13572 MeV	2.014102 u	1.007051 u	1.50346 MeV	0.002388 u	2.22452 MeV	1.11226 MeV
	1	2	14.9498 MeV	3.016049 u	1.005350 u	3.08815 MeV	0.0091058 u	2.22452 MeV	2.8273 MeV
^{3}He	2	1	14.9312 MeV	3.016029 u	1.005343 u	3.09433 MeV	0.0082857 u	7.7181 MeV	2.5727 MeV

In the table above it can be seen that the decay of a neutron, as well as the transformation of tritium into helium-3, releases energy; hence, it manifests a stronger bound new state when measured against the mass of an equal number of neutrons (and also a lighter state per number of total hadrons). Such reactions are not driven by changes in binding energies as calculated from previously fixed N and Z numbers of neutrons and protons, but rather in decreases in the total mass of the nuclide/per nucleon, with the reaction. (Note that the Binding Energy given above for hydrogen-1 is the atomic binding energy, not the nuclear binding energy which would be zero.)

ISOTOPES

Isotopes are variants of a particular chemical element which differ in neutron number, and consequently in nucleon number. All isotopes of a given element have the same number of protons but different numbers of neutrons in each atom.

The meaning behind the name is that different isotopes of a single element occupy the same position on the periodic table. It was coined by a Scottish doctor and writer Margaret Todd in 1913 in a suggestion to chemist Frederick Soddy.

The number of protons within the atom's nucleus is called atomic number and is equal to the number of electrons in the neutral (non-ionized) atom. Each atomic number identifies a specific element, but not the isotope; an atom of a given element may have a wide range in its number of neutrons. The number of nucleons (both protons and neutrons) in the nucleus is the atom's mass number, and each isotope of a given element has a different mass number.

For example, carbon-12, carbon-13, and carbon-14 are three isotopes of the element carbon with mass numbers 12, 13, and 14, respectively. The atomic number of carbon is 6, which means that every carbon atom has 6 protons, so that the neutron numbers of these isotopes are 6, 7, and 8 respectively.

Isotope vs. Nuclide

A nuclide is a species of an atom with a specific number of protons and neutrons in the nucleus, for example carbon-13 with 6 protons and 7 neutrons. The nuclide concept (referring to individual nuclear species) emphasizes nuclear properties over chemical properties, whereas the isotope concept (grouping all atoms of each element) emphasizes chemical over nuclear. The neutron number has large effects on nuclear properties, but its effect on chemical properties is negligible for most elements. Even in the case of the lightest elements where the ratio of neutron number to atomic number varies the most between isotopes it usually has only a small effect, although it does matter in some circumstances (for hydrogen, the lightest element, the isotope effect is large enough to strongly affect biology). The term isotopes (originally also isotopic elements, now sometimes isotopic nuclides) is intended to imply comparison (like synonyms or isomers), for example: the nuclides $_6^{12}C, _6^{13}C, _6^{14}C$ are isotopes (nuclides with the same atomic number but different mass numbers), but $_{18}^{40}Ar, _{19}^{40}K, _{20}^{40}Ca$ are isobars (nuclides with the same mass number). However, because isotope is the older term, it is better known than nuclide, and is still sometimes used in contexts where nuclide might be more appropriate, such as nuclear technology and nuclear medicine.

Notation

An isotope and nuclide is specified by the name of the particular element (this indicates the atomic number) followed by a hyphen and the mass number (e.g. helium-3, helium-4, carbon-12, carbon-14, uranium-235 and uranium-239). When a chemical symbol is used, e.g. "C" for carbon, standard notation (now known as "AZE notation" because A is the mass number, Z the atomic number, and E for element) is to indicate the mass number (number of nucleons) with a superscript at the upper left of the chemical symbol and to indicate the atomic number with a subscript at the lower left (e.g. $^{3}_{2}\text{He}$, $^{4}_{2}\text{He}$, $^{12}_{6}\text{C}$, $^{14}_{6}\text{C}$, and $^{239}_{92}\text{U}$). Because the atomic number is given by the element symbol, it is common to state only the mass number in the superscript and leave out the atomic number subscript (e.g. ^{3}He, ^{4}He, ^{12}C, ^{14}C, ^{235}U). The letter m is sometimes appended after the mass number to indicate a nuclear isomer, a metastable or energetically-excited nuclear state (as opposed to the lowest-energy ground state), for example $^{180m}_{73}\text{Ta}$ (tantalum-180m).

Radioactive, Primordial and Stable Isotopes

Some isotopes/nuclides are radioactive, and are therefore referred to as radioisotopes or radionuclides, whereas others have never been observed to decay radioactively and are referred to as stable isotopes or stable nuclides. For example, ^{14}C is a radioactive form of carbon, whereas ^{12}C and ^{13}C are stable isotopes. There are about 339 naturally occurring nuclides on Earth, of which 286 are primordial nuclides, meaning that they have existed since the Solar System's formation.

Primordial nuclides include 34 nuclides with very long half-lives (over 100 million years) and 252 that are formally considered as "stable nuclides", because they have not been observed to decay. In most cases, for obvious reasons, if an element has stable isotopes, those isotopes predominate in the elemental abundance found on Earth and in the Solar System. However, in the cases of three elements (tellurium, indium, and rhenium) the most abundant isotope found in nature is actually one (or two) extremely long-lived radioisotope(s) of the element, despite these elements having one or more stable isotopes.

Theory predicts that many apparently "stable" isotopes/nuclides are radioactive, with extremely long half-lives (discounting the possibility of proton decay, which would make all nuclides ultimately unstable). Some stable nuclides are in theory energetically susceptible to other known forms of decay, such as alpha decay or double beta decay, but no decay products have yet been observed, and so these isotopes are said to be "observationally stable". The predicted half-lives for these nuclides often greatly exceed the estimated age of the universe, and in fact there are also 31 known radionuclides with half-lives longer than the age of the universe.

Adding in the radioactive nuclides that have been created artificially, there are 3,339 currently known nuclides. These include 905 nuclides that are either stable or have half-lives longer than 60 minutes.

Variation in Properties between Isotopes

Chemical and Molecular Properties

A neutral atom has the same number of electrons as protons. Thus different isotopes of a given element all have the same number of electrons and share a similar electronic structure. Because the chemical behavior of an atom is largely determined by its electronic structure, different isotopes exhibit nearly identical chemical behavior.

The main exception to this is the kinetic isotope effect: due to their larger masses, heavier isotopes tend to react somewhat more slowly than lighter isotopes of the same element. This is most pronounced by far for protium (^{1}H), deuterium (^{2}H), and tritium (^{3}H), because deuterium has twice the mass of protium and tritium has three times the mass of protium. These mass differences also affect the behavior of their respective chemical bonds, by changing the center of gravity (reduced mass) of the atomic systems. However, for heavier elements the relative mass difference between isotopes is much less, so that the mass-difference effects on chemistry are usually negligible. (Heavy elements also have relatively more neutrons than lighter elements, so the ratio of the nuclear mass to the collective electronic mass is slightly greater.)

Similarly, two molecules that differ only in the isotopes of their atoms (isotopologues) have identical electronic structure, and therefore almost indistinguishable physical and chemical properties (again with deuterium and tritium being the primary exceptions). The vibrational modes of a molecule are determined by its shape and by the masses of its constituent atoms; so different isotopologues have different sets of vibrational modes. Because vibrational modes allow a molecule to absorb photons of corresponding energies, isotopologues have different optical properties in the infrared range.

Nuclear Properties and Stability

Atomic nuclei consist of protons and neutrons bound together by the residual strong force. Because protons are positively charged, they repel each other. Neutrons, which are electrically neutral, stabilize the nucleus in two ways. Their copresence pushes protons slightly apart, reducing the electrostatic repulsion between the protons, and they exert the attractive nuclear force on each other and on protons. For this reason, one or more neutrons are necessary for two or more protons to bind into a nucleus. As the number of protons increases, so does the ratio of neutrons to protons necessary to ensure a stable nucleus. For example, although the neutron:proton ratio of $^3_2$He is 1:2, the neutron:proton ratio of $^{238}_{92}$U is greater than 3:2. A number of lighter elements have stable nuclides with the ratio 1:1 (Z = N). The nuclide $^{40}_{20}$Ca (calcium-40) is observationally the heaviest stable nuclide with the same number of neutrons and protons; (theoretically, the heaviest stable one is sulfur-32). All stable nuclides heavier than calcium-40 contain more neutrons than protons.

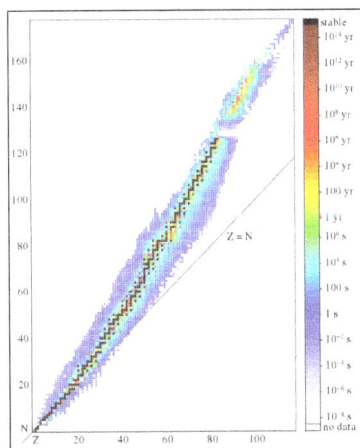

Isotope half-lives. The plot for stable isotopes diverges from the line
Z = N as the element number Z becomes larger.

Numbers of Isotopes Per Element

Of the 80 elements with a stable isotope, the largest number of stable isotopes observed for any element is ten (for the element tin). No element has nine stable isotopes. Xenon is the only element with eight stable isotopes. Four elements have seven stable isotopes, eight have six stable isotopes, ten have five stable isotopes, nine have four stable isotopes, five have three stable isotopes, 16 have two stable isotopes (counting $^{180m}_{73}$Ta as stable), and 26 elements have only a single stable isotope (of these, 19 are so-called mononuclidic elements, having a single primordial stable isotope that dominates and fixes the atomic weight of the natural element to high precision; 3 radioactive mononuclidic elements occur as well). In total, there are 252 nuclides that have not been observed to decay. For the 80 elements that have one or more stable isotopes, the average number of stable isotopes is 252/80 = 3.15 isotopes per element.

Even and Odd Nucleon Numbers

The proton:neutron ratio is not the only factor affecting nuclear stability. It depends also on evenness or oddness of its atomic number Z, neutron number N and, consequently, of their sum, the mass number A. Oddness of both Z and N tends to lower the nuclear binding energy, making odd nuclei, generally, less stable. This remarkable difference of nuclear binding energy between neighbouring nuclei, especially of odd-A isobars, has important consequences: unstable isotopes with a nonoptimal number of neutrons or protons decay by beta decay (including positron decay), electron capture or other exotic means, such as spontaneous fission and cluster decay.

The majority of stable nuclides are even-proton-even-neutron, where all numbers Z, N, and A are even. The odd-A stable nuclides are divided (roughly evenly) into odd-proton-even-neutron, and even-proton-odd-neutron nuclides. Odd-proton-odd-neutron nuclei are the least common.

Table: Even/Odd Z, N (Hydrogen-1 included as OE).

p,n	EE	OO	EO	OE	Total
Stable	146	5	53	48	252
Long-lived	22	4	3	5	34
All primordial	168	9	56	53	286

Even Atomic Number

The 148 even-proton, even-neutron (EE) nuclides comprise ~ 58% of all stable nuclides and all have spin 0 because of pairing. There are also 22 primordial long-lived even-even nuclides. As a result, each of the 41 even-numbered elements from 2 to 82 has at least one stable isotope, and most of these elements have several primordial isotopes. Half of these even-numbered elements have six or more stable isotopes. The extreme stability of helium-4 due to a double pairing of 2 protons and 2 neutrons prevents any nuclides containing five or eight nucleons from existing for long enough to serve as platforms for the buildup of heavier elements via nuclear fusion in stars.

These 53 stable nuclides have an even number of protons and an odd number of neutrons. They are a minority in comparison to the even-even isotopes, which are about 3 times as numerous. Among the 41 even-Z elements that have a stable nuclide, only two elements (argon and cerium) have no even-odd stable nuclides. One element (tin) has three. There are 24 elements that have one even-odd nuclide and 13 that have two odd-even nuclides. Of 35 primordial radionuclides there exist four even-odd nuclides, including the fissile $^{235}_{92}$U Because of their odd neutron numbers, the even-odd nuclides tend to have large neutron capture cross sections, due to the energy that results from neutron-pairing effects. These stable even-proton odd-neutron nuclides tend to be uncommon by abundance in nature, generally because, to form and enter into primordial abundance, they must have escaped capturing neutrons to form yet other stable even-even isotopes, during both the s-process and r-process of neutron capture, during nucleosynthesis in stars. For this reason, only $^{195}_{78}$Pt and $^{9}_{4}$Be re the most naturally abundant isotopes of their element.

Table: Even-odd long-lived.

	Decay	Half-life
$^{113}_{48}$Cd	beta	$7.7 \times 10^{15} a$
$^{147}_{62}$Sm	alpha	$1.06 \times 10^{11} a$
$^{235}_{92}$U	alpha	$7.04 \times 10^{8} a$

Odd Atomic Number

Forty-eight stable odd-proton-even-neutron nuclides, stabilized by their paired neutrons, form most of the stable isotopes of the odd-numbered elements; the very few odd-proton-odd-neutron nuclides comprise the others. There are 41 odd-numbered elements with Z = 1 through 81, of which 39 have stable isotopes (the elements technetium $\left(_{43}Tc\right)$ and promethium $\left(_{61}Pm\right)$ have no stable isotopes). Of these 39 odd Z elements, 30 elements (including hydrogen-1 where 0 neutrons is even) have one stable odd-even isotope, and nine elements: chlorine $\left(_{17}CI\right)$ potassium $\left(_{19}K\right)$, copper $\left(_{29}Cu\right)$, gallium $\left(_{31}Ga\right)$, bromine $\left(_{35}Br\right)$, silver $\left(_{47}Ag\right)$, antimony $\left(_{51}Sb\right)$, iridium $\left(_{77}Ir\right)$, and thallium $\left(_{81}Tl\right)$, have two odd-even stable isotopes each. This makes a total 30 + 2(9) = 48 stable odd-even isotopes.

There are also five primordial long-lived radioactive odd-even isotopes, $^{87}_{37}$Rb, $^{115}_{49}$In, $^{187}_{75}$Re, $^{151}_{63}$Eu, and $^{209}_{83}$Bi. The last two were only recently found to decay, with half-lives greater than 10^{18} years.

Only five stable nuclides contain both an odd number of protons and an odd number of neutrons. The first four "odd-odd" nuclides occur in low mass nuclides, for which changing a proton to a neutron or vice versa would lead to a very lopsided proton-neutron ratio $\left(_{1}^{2}H, _{3}^{6}Li, _{5}^{10}B, \text{ and } _{7}^{14}N; \text{ spin } 1, 1, 3, 1\right)$ The only other entirely "stable" odd-odd nuclide, $^{180m}_{73}$Ta (spin 9), is thought to be the rarest of the 252 stable isotopes, and is the only primordial nuclear isomer, which has not yet been observed to decay despite experimental attempts. Many odd-odd radionuclides (like tantalum-180) with comparatively short half lives are known. Usually, they beta-decay to their nearby even-even isobars that have paired protons and paired neutrons. Of the nine primordial odd-odd nuclides (five stable and four radioactive with long half lives), only $_{7}^{14}N$ is the most common isotope of a common element. This is the case because it is a part of the CNO cycle. The nuclides $_{3}^{6}Li$ and $_{5}^{10}B$ are minority isotopes of elements that are themselves rare compared to other light elements, whereas the other six isotopes make up only a tiny percentage of the natural abundance of their elements.

Odd Neutron Number

Actinides with odd neutron number are generally fissile (with thermal neutrons), whereas those with even neutron number are generally not, though they are fissionable with fast neutrons. All observationally stable odd-odd nuclides have nonzero integer spin. This is because the single unpaired neutron and unpaired proton have a larger nuclear force attraction to each other if their spins are aligned (producing a total spin of at least 1 unit), instead of anti-aligned.

Only $^{195}_{78}$Pt, $^{9}_{4}$Be and $^{14}_{7}$N have odd neutron number and are the most naturally abundant isotope of their element.

Table: Neutron number parity (^{1}H with O neutrons included as even).

N	Even	Odd
Stable	195	58
Long-lived	26	7
All primordial	221	65

Occurrence in Nature

Elements are composed of one nuclide (mononuclidic elements) or of more naturally occurring isotopes. The unstable (radioactive) isotopes are either primordial or postprimordial. Primordial isotopes were a product of stellar nucleosynthesis or another type of nucleosynthesis such as cosmic ray spallation, and have persisted down to the present because their rate of decay is so slow (e.g., uranium-238 and potassium-40). Post-primordial isotopes were created by cosmic ray bombardment as cosmogenic nuclides (e.g., tritium, carbon-14), or by the decay of a radioactive primordial isotope to a radioactive radiogenic nuclide daughter (e.g., uranium to radium). A few isotopes are naturally synthesized as nucleogenic nuclides, by some other natural nuclear reaction, such as when neutrons from natural nuclear fission are absorbed by another atom.

As discussed above, only 80 elements have any stable isotopes, and 26 of these have only one stable isotope. Thus, about two-thirds of stable elements occur naturally on Earth in multiple stable isotopes, with the largest number of stable isotopes for an element being ten, for tin ($_{50}$Sn). There are about 94 elements found naturally on Earth (up to plutonium inclusive), though some are detected only in very tiny amounts, such as plutonium-244. Scientists estimate that the elements that occur naturally on Earth (some only as radioisotopes) occur as 339 isotopes (nuclides) in total. Only 252 of these naturally occurring nuclides are stable in the sense of never having been observed to decay as of the present time. An additional 34 primordial nuclides (to a total of 286 primordial nuclides), are radioactive with known half-lives, but have half-lives longer than 100 million years, allowing them to exist from the beginning of the Solar System.

All the known stable nuclides occur naturally on Earth; the other naturally occurring nuclides are radioactive but occur on Earth due to their relatively long half-lives, or else due to other means of ongoing natural production. These include the afore-mentioned cosmogenic nuclides, the nucleogenic nuclides, and any radiogenic nuclides formed by ongoing decay of a primordial radioactive nuclide, such as radon and radium from uranium.

An additional ~ 3000 radioactive nuclides not found in nature have been created in nuclear reactors and in particle accelerators. Many short-lived nuclides not found naturally on Earth have also been observed by spectroscopic analysis, being naturally created in stars or supernovae. An example is aluminium-26, which is not naturally found on Earth, but is found in abundance on an astronomical scale.

The tabulated atomic masses of elements are averages that account for the presence of multiple isotopes with different masses. Before the discovery of isotopes, empirically determined noninteger values of atomic mass confounded scientists. For example, a sample of chlorine contains 75.8% chlorine-35 and 24.2% chlorine-37, giving an average atomic mass of 35.5 atomic mass units.

According to generally accepted cosmology theory, only isotopes of hydrogen and helium, traces of some isotopes of lithium and beryllium, and perhaps some boron, were created at the Big Bang, while all other nuclides were synthesized later, in stars and supernovae, and in interactions between energetic particles such as cosmic rays, and previously produced nuclides. The respective abundances of isotopes on Earth result from the quantities formed by these processes, their spread through the galaxy, and the rates of decay for isotopes that are unstable. After the initial coalescence of the Solar System, isotopes were redistributed according to mass, and the isotopic composition of elements varies slightly from planet to planet. This sometimes makes it possible to trace the origin of meteorites.

Atomic Mass of Isotopes

The atomic mass (m_r) of an isotope (nuclide) is determined mainly by its mass number (i.e. number of nucleons in its nucleus). Small corrections are due to the binding energy of the nucleus the slight difference in mass between proton and neutron, and the mass of the electrons associated with the atom, the latter because the electron:nucleon ratio differs among isotopes.

The mass number is a dimensionless quantity. The atomic mass, on the other hand, is measured using the atomic mass unit based on the mass of the carbon-12 atom. It is denoted with symbols "u" (for unified atomic mass unit) or "Da" (for dalton).

The atomic masses of naturally occurring isotopes of an element determine the atomic mass of the element. When the element contains N isotopes, the expression below is applied for the average atomic mass $\overline{}$:

$$\overline{m}_a = m_1 x_1 + m_2 x_2 + \ldots + m_N x_N$$

where m_1, m_2, ..., m_N are the atomic masses of each individual isotope, and x_1, ..., x_N are the relative abundances of these isotopes.

Applications of Isotopes

Purification of Isotopes

Several applications exist that capitalize on properties of the various isotopes of a given element. Isotope separation is a significant technological challenge, particularly with heavy elements such as uranium or plutonium. Lighter elements such as lithium, carbon, nitrogen, and oxygen are commonly separated by gas diffusion of their compounds such as CO and NO. The separation of hydrogen and deuterium is unusual because it is based on chemical rather than physical properties, for example in the Girdler sulfide process. Uranium isotopes have been separated in bulk by gas diffusion, gas centrifugation, laser ionization separation, and (in the Manhattan Project) by a type of production mass spectrometry.

Use of Chemical and Biological Properties

- Isotope analysis is the determination of isotopic signature, the relative abundances of isotopes of a given element in a particular sample. For biogenic substances in particular, significant variations of isotopes of C, N and O can occur. Analysis of such variations has a wide range of applications, such as the detection of adulteration in food products or the geographic origins of products using isoscapes. The identification of certain meteorites as having originated on Mars is based in part upon the isotopic signature of trace gases contained in them.

- Isotopic substitution can be used to determine the mechanism of a chemical reaction via the kinetic isotope effect.

- Another common application is isotopic labeling, the use of unusual isotopes as tracers or markers in chemical reactions. Normally, atoms of a given element are indistinguishable from each other. However, by using isotopes of different masses, even different nonradioactive stable isotopes can be distinguished by mass spectrometry or infrared spectroscopy. For example, in 'stable isotope labeling with amino acids in cell culture (SILAC)' stable isotopes are used to quantify proteins. If radioactive isotopes are used, they can be detected by the radiation they emit (this is called radioisotopic labeling).

- Isotopes are commonly used to determine the concentration of various elements or substances using the isotope dilution method, whereby known amounts of isotopically-substituted compounds are mixed with the samples and the isotopic signatures of the resulting mixtures are determined with mass spectrometry.

Use of Nuclear Properties

- A technique similar to radioisotopic labeling is radiometric dating: using the known half-life of an unstable element, one can calculate the amount of time that has elapsed since a known concentration of isotope existed. The most widely known example is radiocarbon dating used to determine the age of carbonaceous materials.

- Several forms of spectroscopy rely on the unique nuclear properties of specific isotopes, both radioactive and stable. For example, nuclear magnetic resonance (NMR) spectroscopy can be used only for isotopes with a nonzero nuclear spin. The most common nuclides used with NMR spectroscopy are ^{1}H, ^{2}D, ^{15}N, ^{13}C, and ^{31}P.

- Mössbauer spectroscopy also relies on the nuclear transitions of specific isotopes, such as ^{57}Fe.

- Radionuclides also have important uses. Nuclear power and nuclear weapons development require relatively large quantities of specific isotopes. Nuclear medicine and radiation oncology utilize radioisotopes respectively for medical diagnosis and treatment.

ISOBARS

Isobars are atoms (nuclides) of different chemical elements that have the same number of nucleons. Correspondingly, isobars differ in atomic number (or number of protons) but have the same mass number. An example of a series of isobars would be ^{40}S, ^{40}Cl, ^{40}Ar, ^{40}K, and ^{40}Ca. The nuclei of these nuclides all contain 40 nucleons; however, they contain varying numbers of protons and neutrons.

The term "isobars" (originally "isobares") for nuclides was suggested by Alfred Walter Stewart in 1918. It is derived from the Greek word isos, meaning "equal" and baros, meaning "weight".

Mass

The same mass number implies neither the same mass of nuclei, nor equal atomic masses of corresponding nuclides. From the Weizsäcker's formula for the mass of a nucleus:

$$m(A,Z) = Zm_p + Nm_n - a_V A + a_S A^{2/3} + a_C \frac{Z^2}{A^{1/3}} + a_A \frac{(N-Z)^2}{A} - \delta(A,Z)$$

where mass number A equals to the sum of atomic number Z and number of neutrons N, and m_p, m_n, a_V, a_S, a_C, a_A are constants, one can see that the mass depends on Z and N non-linearly, even for a constant mass number. For odd A, it is admitted that $\delta = 0$ and the mass dependence on Z is convex (or on N or N − Z, it does not matter for a constant A). This explains that beta-decay is energetically favorable for neutron-rich nuclides, and positron decay is favorable for strongly neutron-deficient nuclides. Both decay modes do not change the mass number, hence an original nucleus and its daughter nucleus are isobars. In both aforementioned cases, a heavier nucleus decays to its lighter isobar.

For even A the δ term has the form:

$$\delta(A,Z) = (-1)^Z a_P A^{-\frac{1}{2}}$$

where aP is another constant. This term, subtracted from the mass expression above, is positive for even-even nuclei and negative for odd-odd nuclei. This means that even-even nuclei, which have not a strong neutron excess or neutron deficiency, have higher binding energy than their odd-odd isobar neighbors. It implies that even-even nuclei are (relatively) lighter and more stable. The difference is especially strong for small A. This effect is also predicted (qualitatively) by other nuclear models and has important consequences.

Stability

The Mattauch isobar rule states that if two adjacent elements on the periodic table have isotopes of the same mass number, (at least) one of these isobars must be a radionuclide (radioactive). In cases of three isobars of sequential elements where the first and last are stable (this is often the case for even-even nuclides,), branched decay of the middle isobar may occur; e.g. radioactive iodine-126 has an almost equal probabilities for two decay modes, which lead to different daughter isotopes: tellurium-126 and xenon-126.

No observationally stable isobars exist for mass numbers 5 (decays to helium-4 plus a proton or neutron), 8 (decays to two helium-4 nuclei), 147, 151, as well as for 209 and above. Two observationally stable isobars exist for 36, 40, 46, 50, 54, 58, 64, 70, 74, 80, 84, 86, 92, 94, 96, 98, 102, 104, 106, 108, 110, 112, 114, 120, 122, 123, 124, 126, 132, 134, 136, 138, 142, 154, 156, 158, 160, 162, 164, 168, 170, 176, 180, 184, 192, 196, 198 and 204.

In theory, no two stable nuclides have the same mass number (since no two nuclides have the same mass number are both stable to beta decay and double beta decay), and no stable nuclides exist for mass number 5, 8, 143-155, 160-162, and $\geq$ 165, since in theory, the beta-decay stable nuclides for these mass number can undergo alpha decay.

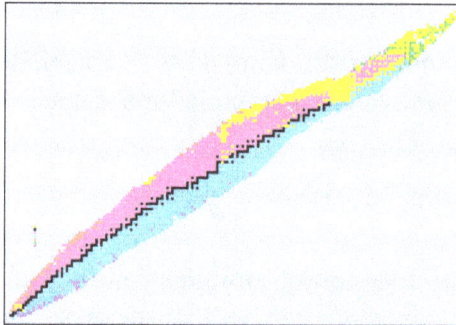

In this chart of nuclides, isobars occur along diagonal lines running from the lower right to upper left. The line of beta-stability includes the observationally stable nuclides shown in black; disconnected 'islands' are a consequence of the Mattauch isobar rule.

HYPERFINE STRUCTURE

In atomic physics, hyperfine structure refers to small shifts and splittings in the energy levels of atoms, molecules, and ions, due to interaction between the state of the nucleus and the state of the electron clouds.

In atoms, hyperfine structure arises from the energy of the nuclear magnetic dipole moment interacting with the magnetic field generated by the electrons and the energy of the nuclear electric quadrupole moment in the electric field gradient due to the distribution of charge within the atom. Molecular hyperfine structure is generally dominated by these two effects, but also includes the energy associated with the interaction between the magnetic moments associated with different magnetic nuclei in a molecule, as well as between the nuclear magnetic moments and the magnetic field generated by the rotation of the molecule.

Hyperfine structure contrasts with fine structure, which results from the interaction between the magnetic moments associated with electron spin and the electrons' orbital angular momentum. Hyperfine structure, with energy shifts typically orders of magnitudes smaller than those of a fine-structure shift, results from the interactions of the nucleus (or nuclei, in molecules) with internally generated electric and magnetic fields.

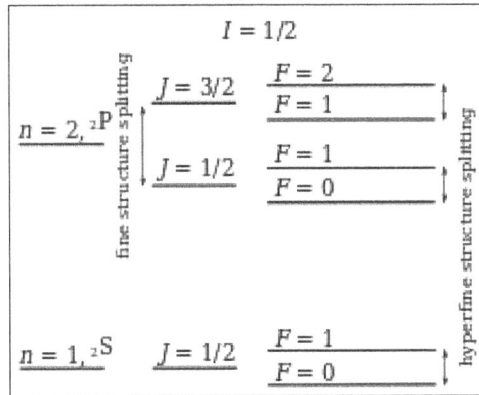

Schematic illustration of fine and hyperfine structure in a neutral hydrogen atom.

Theory

The theory of hyperfine structure comes directly from electromagnetism, consisting of the interaction of the nuclear multipole moments (excluding the electric monopole) with internally generated fields. The theory is derived first for the atomic case, but can be applied to each nucleus in a molecule. Following this there is a discussion of the additional effects unique to the molecular case.

Atomic Hyperfine Structure

Magnetic Dipole

The dominant term in the hyperfine Hamiltonian is typically the magnetic dipole term. Atomic nuclei with a non-zero nuclear spin I have a magnetic dipole moment, given by:

$$\mu_I = g_I \mu_N \mathbf{I},$$

where g_I is the g-factor and μ_N is the nuclear magneton.

There is an energy associated with a magnetic dipole moment in the presence of a magnetic field. For a nuclear magnetic dipole moment, μI, placed in a magnetic field, B, the relevant term in the Hamiltonian is given by:

$$\hat{H}_D = -\hat{\boldsymbol{\mu}}_I \cdot \mathbf{B}.$$

In the absence of an externally applied field, the magnetic field experienced by the nucleus is that associated with the orbital (l) and spin (s) angular momentum of the electrons:

$$\mathbf{B} \equiv \mathbf{B}_{el} = \mathbf{B}_{el}^{l} + \mathbf{B}_{el}^{s}.$$

Electron orbital angular momentum results from the motion of the electron about some fixed external point that we shall take to be the location of the nucleus. The magnetic field at the nucleus due to the motion of a single electron, with charge -e at a position r relative to the nucleus, is given by:

$$\mathbf{B}_{el}^{l} = \frac{\mu_0}{4\pi} \frac{-e\mathbf{v} \times -\mathbf{r}}{r^3},$$

where $-r$ gives the position of the nucleus relative to the electron. Written in terms of the Bohr magneton, this gives:

$$\mathbf{B}_{el}^{l} = -2\mu_B \frac{\mu_0}{4\pi} \frac{1}{r^3} \frac{\mathbf{r} \times m_e \mathbf{v}}{\hbar}.$$

Recognizing that mev is the electron momentum, p, and that $r \times p / \hbar$ is the orbital angular momentum in units of $\hbar$, l, we can write:

$$B_{el}^{l} = -2\mu_B \frac{\mu_0}{4\pi} \frac{1}{r^3} \mathbf{l}.$$

For a many electron atom this expression is generally written in terms of the total orbital angular momentum, L , by summing over the electrons and using the projection operator, ϕ_i^{l}, where $\sum_i \mathbf{l}_i = \sum_i \phi_i^{l} \mathbf{L}$. For states with a well defined projection of the orbital angular momentum, L_z, we can write $\phi_i^{l} = \hat{l}_{z_i} / L_z$, giving:

$$\mathbf{B}_{el}^{l} = -2\mu_B \frac{\mu_0}{4\pi} \frac{1}{L_z} \sum_i \frac{\hat{l}_{zi}}{r_i^3} \mathbf{L}.$$

The electron spin angular momentum is a fundamentally different property that is intrinsic to the particle and therefore does not depend on the motion of the electron. Nonetheless it is angular momentum and any angular momentum associated with a charged particle results in a magnetic dipole moment, which is the source of a magnetic field. An electron with spin angular momentum, s, has a magnetic moment, μs , given by:

$$\mu_s = -g_s \mu_B \mathbf{s},$$

where gs is the electron spin g-factor and the negative sign is because the electron is negatively charged (consider that negatively and positively charged particles with identical mass, travelling on equivalent paths, would have the same angular momentum, but would result in currents in the opposite direction).

The magnetic field of a dipole moment, μs, is given by:

$$\mathbf{B}^s_{el} = \frac{\mu_0}{4\pi r^3}\left(3(\mu_s \cdot \hat{\mathbf{r}})\hat{\mathbf{r}} - \mu_s\right) + \frac{2\mu_0}{3}\mu_s \delta^3(\mathbf{r}).$$

The complete magnetic dipole contribution to the hyperfine Hamiltonian is thus given by:

$$\hat{H}_D = 2g_I\mu_N\mu_B\frac{\mu_0}{4\pi}\frac{1}{L_z}\sum_i \frac{\hat{l}_{zi}}{r_i^3}\mathbf{I}\cdot\mathbf{L}$$

$$+ g_I\mu_N g_s\mu_B\frac{\mu_0}{4\pi}\frac{1}{S_z}\sum_i \frac{\hat{s}_{zi}}{r_i^3}\left\{3(\mathbf{I}\cdot\hat{\mathbf{r}})(\mathbf{S}\cdot\hat{\mathbf{r}}) - \mathbf{I}\cdot\mathbf{S}\right\}$$

$$+ \frac{2}{3}g_I\mu_N g_s\mu_B\mu_0\frac{1}{S_z}\sum_i \hat{s}_{zi}\delta^3(\mathbf{r}_i)\mathbf{I}\cdot\mathbf{S}.$$

The first term gives the energy of the nuclear dipole in the field due to the electronic orbital angular momentum. The second term gives the energy of the "finite distance" interaction of the nuclear dipole with the field due to the electron spin magnetic moments. The final term, often known as the "Fermi contact" term relates to the direct interaction of the nuclear dipole with the spin dipoles and is only non-zero for states with a finite electron spin density at the position of the nucleus (those with unpaired electrons in s-subshells). It has been argued that one may get a different expression when taking into account the detailed nuclear magnetic moment distribution.

For states with $l \neq 0$ this can be expressed in the form:

$$\hat{H}_D = 2g_I\mu_B\mu_N\frac{\mu_0}{4\pi}\frac{\mathbf{I}\cdot\mathbf{N}}{r^3},$$

where:

$$\mathbf{N} = \mathbf{l} - (g_s/2)\left[\mathbf{s} - 3(\mathbf{s}\cdot\hat{\mathbf{r}})\hat{\mathbf{r}}\right].$$

If hyperfine structure is small compared with the fine structure (sometimes called IJ-coupling by analogy with LS-coupling), I and J are good quantum numbers and matrix elements of $\hat{H}_D$ can be approximated as diagonal in I and J. In this case (generally true for light elements), we can project N onto J (where J = L ı S is the total electronic angular momentum) and we have:

$$\hat{H}_D = 2g_I\mu_B\mu_N\frac{\mu_0}{4\pi}\frac{\mathbf{N}\cdot\mathbf{J}}{\mathbf{J}\cdot\mathbf{J}}\frac{\mathbf{I}\cdot\mathbf{J}}{r^3}.$$

This is commonly written as:

$$\hat{H}_D = \hat{A}\mathbf{I}\cdot\mathbf{J},$$

with $\langle \hat{A} \rangle$ being the hyperfine structure constant which is determined by experiment. Since I. J = ½ {F.F - I.I - J.J} (where F = I + J is the total angular momentum), this gives an energy of:

$$\Delta E_{\mathrm{D}} = \frac{1}{2}\langle \hat{A}\rangle[F(F+1)-I(I+1)-J(J+1)].$$

In this case the hyperfine interaction satisfies the Landé interval rule.

Electric Quadrupole

Atomic nuclei with spin $I \geq 1$ have an electric quadrupole moment. In the general case this is represented by a rank-2 tensor, $\underline{\underline{Q}}$, with components given by:

$$Q_{ij} = \frac{1}{e}\int \left(3x_i'x_j' - (r')^2\delta_{ij}\right)\rho(\mathbf{r}')d^3r',$$

where i and j are the tensor indices running from 1 to 3, x_i and x_j are the spatial variables x, y and z depending on the values of i and j respectively, δij is the Kronecker delta and $\rho(r)$ is the charge density. Being a 3-dimensional rank-2 tensor, the quadrupole moment has 3² = 9 components. From the definition of the components it is clear that the quadrupole tensor is a symmetric matrix ($Q_{ij} = Q_{ji}$) that is also traceless ($\Sigma_i Q_{ii} = 0$), giving only five components in the irreducible representation. Expressed using the notation of irreducible spherical tensors we have:

$$T_m^2(Q) = \sqrt{\frac{4\pi}{5}}\int \rho(\mathbf{r}')(r')^2 Y_m^2(\theta',\phi')d^3r'.$$

The energy associated with an electric quadrupole moment in an electric field depends not on the field strength, but on the electric field gradient, confusingly labelled q, another rank-2 tensor given by the outer product of the del operator with the electric field vector:

$$\underline{\underline{q}} = \nabla \otimes \mathbf{E},$$

with components given by:

$$q_{ij} = \frac{\partial^2 V}{\partial x_i \partial x_j}.$$

Again it is clear this is a symmetric matrix and, because the source of the electric field at the nucleus is a charge distribution entirely outside the nucleus, this can be expressed as a 5-component spherical tensor, $T^2(q)$, with:

$$T_0^2(q) = \frac{\sqrt{6}}{2}q_{zz}$$

$$T_{+1}^2(q) = -q_{xz} - iq_{yz}$$

$$T^2_{+2}(q) = \frac{1}{2}(q_{xx} - q_{yy}) + iq_{xy},$$

where:

$$T^2_{-m}(q) = (-1)^m T^2_{+m}(q)^*.$$

The quadrupolar term in the Hamiltonian is thus given by:

$$\hat{H}_Q = -eT^2(Q) \cdot T^2(q) = -e\sum_m (-1)^m T^2_m(Q) T^2_{-m}(q).$$

A typical atomic nucleus closely approximates cylindrical symmetry and therefore all off-diagonal elements are close to zero. For this reason the nuclear electric quadrupole moment is often represented by Q_{zz}.

Molecular Hyperfine Structure

The molecular hyperfine Hamiltonian includes those terms already derived for the atomic case with a magnetic dipole term for each nucleus with $I > 0$ and an electric quadrupole term for each nucleus with $I \geq 1$ The magnetic dipole terms were first derived for diatomic molecules by Frosch and Foley, and the resulting hyperfine parameters are often called the Frosch and Foley parameters.

In addition to the effects described above, there are a number of effects specific to the molecular case.

Direct Nuclear Spin–spin

Each nucleus with $I > 0$ has a non-zero magnetic moment that is both the source of a magnetic field and has an associated energy due to the presence of the combined field of all of the other nuclear magnetic moments. A summation over each magnetic moment dotted with the field due to each other magnetic moment gives the direct nuclear spin–spin term in the hyperfine Hamiltonian,

$$\hat{H}_{II}.$$

$$\hat{H}_{II} = -\sum_{\alpha \neq \alpha'} \hat{\imath}_\alpha \cdot \mathbf{B}_{\alpha'},$$

where α and α' are indices representing the nucleus contributing to the energy and the nucleus that is the source of the field respectively. Substituting in the expressions for the dipole moment in terms of the nuclear angular momentum and the magnetic field of a dipole, both given above, we have:

$$\hat{H}_{II} = \frac{\mu_0 \mu_N^2}{4\pi} \sum_{\alpha \neq \alpha'} \frac{g_\alpha g_{\alpha'}}{R^3_{\alpha\alpha'}} \{ \mathbf{I}_\alpha \cdot \mathbf{I}_{\alpha'} - 3(\mathbf{I}_\alpha \cdot \hat{\mathbf{R}}_{\alpha\alpha'})(\mathbf{I}_{\alpha'} \cdot \hat{\mathbf{R}}_{\alpha\alpha'}) \}.$$

Nuclear Spin–rotation

The nuclear magnetic moments in a molecule exist in a magnetic field due to the angular momentum, T (R is the internuclear displacement vector), associated with the bulk rotation of the molecule, thus:

$$\hat{H}_{IR} = \frac{e\mu_0\mu_N\hbar}{4\pi} \sum_{\alpha \neq \alpha'} \frac{1}{R_{\alpha\alpha'}^3} \left\{ \frac{Z_\alpha g_{\alpha'}}{M_\alpha} \mathbf{I}_{\alpha'} + \frac{Z_{\alpha'} g_\alpha}{M_{\alpha'}} \mathbf{I}_\alpha \right\} \cdot \mathbf{T}.$$

Small Molecule Hyperfine Structure

A typical simple example of the hyperfine structure due to the interactions discussed above is in the rotational transitions of hydrogen cyanide ($^1H^{12}C^{14}N$) in its ground vibrational state. Here, the electric quadrupole interaction is due to the ^{14}N-nucleus, the hyperfine nuclear spin-spin splitting is from the magnetic coupling between nitrogen, ^{14}N (I_N=1), and hydrogen, ^1H (I_H=1/2), and a hydrogen spin-rotation interaction due to the 1H-nucleus. These contributing interactions to the hyperfine structure in the molecule are listed here in descending order of influence. Sub-doppler techniques have been used to discern the hyperfine structure in HCN rotational transitions .

The dipole selection rules for HCN hyperfine structure transitions are $\Delta J = 1$, $\Delta F = 0, \pm 1$ where J is the rotational quantum number and F is the total rotational quantum number inclusive of nuclear spin $F = J + I_N$ respectively. The lowest transition $J = 1 \rightarrow 0$ splits into a hyperfine triplet. Using the selection rules, the hyperfine pattern of $J = 2 \rightarrow 1$ transition and higher dipole transitions is in the form of a hyperfine sextet. However, one of these components $\Delta F = -1$ carries only 0.6% of the rotational transition intensity in the case of $J = 2 \rightarrow 1$. This contribution drops for increasing J. So, from $J = 2 \rightarrow 1$ upwards the hyperfine pattern consists of three very closely spaced stronger hyperfine components $\left(\Delta J = 1, \Delta F = 1 \right)$ together with two widely spaced components; one on the low frequency side and one on the high frequency side relative to the central hyperfine triplet. Each of these outliers carry $\sim \frac{1}{2} J^2$ (J is the upper rotational quantum number of the allowed dipole transition) the intensity of the entire transition. For consecutively higher-J transitions, there are small but significant changes in the relative intensities and positions of each individual hyperfine component.

Measurements

Hyperfine interactions can be measured, among other ways, in atomic and molecular spectra and in electron paramagnetic resonance spectra of free radicals and transition-metal ions.

Applications

Astrophysics

As the hyperfine splitting is very small, the transition frequencies are usually not located in the optical, but are in the range of radio- or microwave (also called sub-millimeter) frequencies.

Hyperfine structure gives the 21 cm line observed in H I regions in interstellar medium.

Carl Sagan and Frank Drake considered the hyperfine transition of hydrogen to be a sufficiently universal phenomenon so as to be used as a base unit of time and length on the Pioneer plaque and later Voyager Golden Record.

In submillimeter astronomy, heterodyne receivers are widely used in detecting electromagnetic signals from celestial objects such as star-forming core or young stellar objects. The separations among neighboring components in a hyperfine spectrum of an observed rotational transition are usually small enough to fit within the receiver's IF band. Since the optical depth varies with frequency, strength ratios among the hyperfine components differ from that of their intrinsic (or optically thin) intensities (these are so-called hyperfine anomalies, often observed in the rotational transitions of HCN). Thus, a more accurate determination of the optical depth is possible. From this we can derive the object's physical parameters.

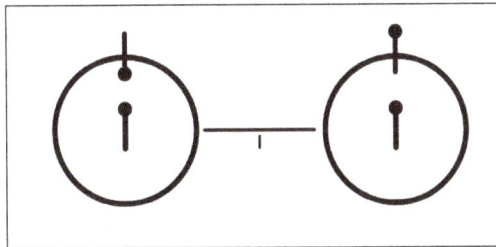

The hyperfine transition as depicted on the Pioneer plaque.

Nuclear Technology

The atomic vapor laser isotope separation (AVLIS) process uses the hyperfine splitting between optical transitions in uranium-235 and uranium-238 to selectively photo-ionize only the uranium-235 atoms and then separate the ionized particles from the non-ionized ones. Precisely tuned dye lasers are used as the sources of the necessary exact wavelength radiation.

Use in Defining the SI Second and Meter

The hyperfine structure transition can be used to make a microwave notch filter with very high stability, repeatability and Q factor, which can thus be used as a basis for very precise atomic clocks. Typically, the hyperfine structure transition frequency of a particular isotope of caesium or rubidium atoms is used as a basis for these clocks.

Due to the accuracy of hyperfine structure transition-based atomic clocks, they are now used as the basis for the definition of the second. One second is now defined to be exactly 9,192,631,770 cycles of the hyperfine structure transition frequency of caesium-133 atoms.

On October 21, 1983, the 17th CCPM defined the metre as the length of the path travelled by light in a vacuum during a time interval of $\dfrac{1}{299,792,458}$ of a second.

Precision Tests of Quantum Electrodynamics

The hyperfine splitting in hydrogen and in muonium have been used to measure the value of the fine structure constant α. Comparison with measurements of α in other physical systems provides a stringent test of QED.

Qubit in Ion-trap Quantum Computing

The hyperfine states of a trapped ion are commonly used for storing qubits in ion-trap quantum computing. They have the advantage of having very long lifetimes, experimentally exceeding ~ 10 min (compared to $\sim 1\,s$ for metastable electronic levels).

The frequency associated with the states' energy separation is in the microwave region, making it possible to drive hyperfine transitions using microwave radiation. However, at present no emitter is available that can be focused to address a particular ion from a sequence. Instead, a pair of laser pulses can be used to drive the transition, by having their frequency difference (detuning) equal to the required transition's frequency. This is essentially a stimulated Raman transition. In addition, near-field gradients have been exploited to individually address two ions separated by approximately 4.3 micrometers directly with microwave radiation.

ATOMIC SPECTROSCOPY

Atomic spectroscopy is the study of the electromagnetic radiation absorbed and emitted by atoms. Since unique elements have characteristic (signature) spectra, atomic spectroscopy, specifically the electromagnetic spectrum or mass spectrum, is applied for determination of elemental compositions. It can be divided by atomization source or by the type of spectroscopy used. In the latter case, the main division is between optical and mass spectrometry. Mass spectrometry generally gives significantly better analytical performance, but is also significantly more complex. This complexity translates into higher purchase costs, higher operational costs, more operator training, and a greater number of components that can potentially fail. Because optical spectroscopy is often less expensive and has performance adequate for many tasks, it is far more common. Atomic absorption spectrometers are one of the most commonly sold and used analytical devices.

Optical Spectroscopy

Electrons exist in energy levels (i.e., atomic orbitals) within an atom. Atomic orbitals are quantized, meaning they exist as defined values instead of being continuous. Electrons may move between orbitals, but in doing so they must absorb or emit energy equal to the energy difference between their atom's specific quantized orbital energy levels. In optical spectroscopy, energy absorbed to move an electron to a higher energy level (higher orbital) and the energy emitted as the electron moves to a lower energy level is absorbed or emitted in the form of photons (light particles). Because each element has a unique number of electrons, an atom will absorb/release energy in a pattern unique to its elemental identity (e.g., Ca, Na, etc.) and thus will absorb/emit photons in a correspondingly unique pattern. The type of atoms present in a sample, or the amount of atoms present in a sample can be deduced from measuring these changes in light wavelength and light intensity.

Optical spectroscopy is further divided into atomic absorption spectroscopy and atomic emission spectroscopy. In atomic absorption spectroscopy, light of a predetermined wavelength is passed through a collection of atoms. If the wavelength of the source light has energy corresponding to the energy difference between two energy levels in the atoms, a portion of the light will be absorbed. The difference between the intensity of the light emitted from the source (e.g., lamp) and the light collected by the detector yields an absorbance value. This absorbance value can then be used to determine the concentration of a given element (or atoms) within the sample. The relationship between the concentration of atoms, the distance the light travels through the collection of atoms, and the portion of the light absorbed is given by the Beer–Lambert law. In atomic emission spectroscopy, the intensity of the emitted light is directly proportional to the concentration of atoms.

Mass Spectrometry

Atomic mass spectrometry is similar to other types of mass spectrometry in that it consists of an ion source, a mass analyzer, and a detector. Atoms' identities are determined by their mass-to-charge ratio (via the mass analyzer) and their concentrations are determined by the number of ions detected. Although considerable research has gone into customizing mass spectrometers for atomic ion sources, it is the ion source that differs most from other forms of mass spectrometry. These ion sources must also atomize samples, or an atomization step must take place before ionization. Atomic ion sources are generally modifications of atomic optical spectroscopy atom sources.

Ion and Atom Sources

Sources can be adapted in many ways, but the lists below give the general uses of a number of sources. Of these, flames are the most common due to their low cost and their simplicity. Although significantly less common, inductively-coupled plasmas, especially when used with mass spectrometers, are recognized for their outstanding analytical performance and their versatility.

For all atomic spectroscopy, a sample must be vaporized and atomized. For atomic mass spectrometry, a sample must also be ionized. Vaporization, atomization, and ionization are often, but not always, accomplished with a single source. Alternatively, one source may be used to vaporize a sample while another is used to atomize (and possibly ionize). An example of this is laser ablation inductively-coupled plasma atomic emission spectrometry, where a laser is used to vaporize a solid sample and an inductively-coupled plasma is used to atomize the vapor.

With the exception of flames and graphite furnaces, which are most commonly used for atomic absorption spectroscopy, most sources are used for atomic emission spectroscopy.

Liquid-sampling sources include flames and sparks (atom source), inductively-coupled plasma (atom and ion source), graphite furnace (atom source), microwave plasma (atom and ion source), and direct-current plasma (atom and ion source). Solid-sampling sources include lasers (atom and vapor source), glow discharge (atom and ion source), arc (atom and ion source), spark (atom and ion source), and graphite furnace (atom and vapor source). Gas-sampling sources include flame (atom source), inductively-coupled plasma (atom and ion source), microwave plasma (atom and ion source), direct-current plasma (atom and ion source), and glow discharge (atom and ion source).

References

- Atomic-physics, science: britannica.com, Retrieved 21 July, 2019

- Arabatzis, t. (2006). Representing electrons: a biographical approach to theoretical entities. University of chicago press. Pp. 70–74, 96. Isbn 978-0-226-02421-9

- Charge-to-mass-ratio, chemistry: byjus.com, Retrieved 7 May, 2019

- Buchwald, j.z.; warwick, a. (2001). Histories of the electron: the birth of microphysics. Mit press. Pp. 195–203. Isbn 978-0-262-52424-7

- Atomic-models: wikilectures.eu, Retrieved 18 April, 2019

- Curtis, l.j. (2003). Atomic structure and lifetimes: a conceptual approach. Cambridge university press. P. 74. Isbn 978-0-521-53635-6

- Daltons-atomic-model: brilliant.org, Retrieved 14 June, 2019

- Dr. Rod nave of the department of physics and astronomy, dr. Rod nave (july 2010). "nuclear binding energy". Hyperphysics - a free web resource from gsu. Georgia state university. Retrieved 2010-07-11

- Thomsons-model, chemistry: byjus.com, Retrieved 28 February, 2019

- Rutherfords-model-of-an-atom, structure-of-atom, chemistry, guides: toppr.com, Retrieved 19 May, 2019

- "nuclear binding energy". How to solve for nuclear binding energy. Guides to solving many of the types of quantitative problems found in chemistry 116. Purdue university. July 2010. Retrieved 2010-07-10. Guides

Interactions of Atoms

Depending upon the frequency and wavelength, an atom shows several types of behavioral aspects when exposed to different radiations. Some of the phenomena which are considered during this exposure are ground state of atom, excited state of atom, emission and absorption spectra. These phenomena elaborated in this chapter will help in gaining a better perspective about the subject.

GROUND STATE

The ground state of an atomic nucleus, atom, or molecule is its lowest energy state. Higher energy states are described as excited states.

The ground state applies to any quantized property of a particle. In chemistry,

- Electron ground states.
- Vibrational ground states.
- Rotational ground states.

of atoms and molecules are important, as are their excited states.

At room temperature, most molecules are in electron and vibrational ground states; higher temperatures are needed for molecules to enter excited states. However, most molecules are in an excited rotational state at room temperature, because less energy is needed for a molecule to enter a rotational excited state than an electron or vibrational excited state.

Electron Ground States

The diagrams below show electron energy levels for a hydrogen atom. In the first diagram, hydrogen is in its electron ground state.

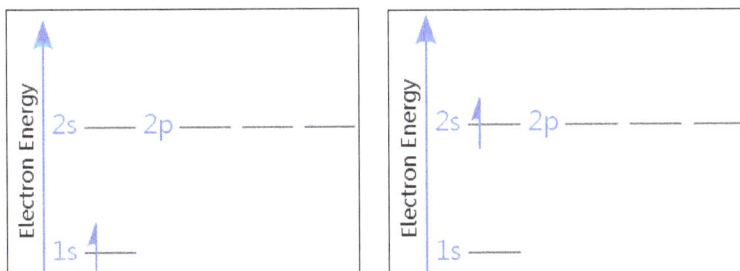

In the second diagram, hydrogen's electron is in a higher energy shell: Hydrogen is no longer in its electron ground state; it is in an excited state. (Remember hydrogen, with one electron, has degenerate subshells, so 2s is degenerate with 2p - in all other atoms, electrons in the 2p subshell have higher energy than in the 2s.

If hydrogen's electron gained more energy it could enter higher sublevels such as 3s, 3p, 4s, 3d, etc. In all of these cases, it would be in an excited state.

Returning to the Ground State

Particles in excited states can return to the ground state by releasing energy. The energy is lost in the form of electromagnetic radiation.

- When electrons in atoms or molecules emit energy and fall into lower energy sublevels, including into the ground state, the energy takes the form of visible or ultraviolet light.

- Vibrational energy is lost by the emission of infrared light.

- Rotational energy is lost by the emission of microwave or far infrared radiation.

Degenerate Ground States

If there is more than one ground state, it means there is more than one lowest energy state.

For example, the diagrams below show two equal energy ground states for hydrogen. In one, the electron spin is + ½; in the other it is -½.

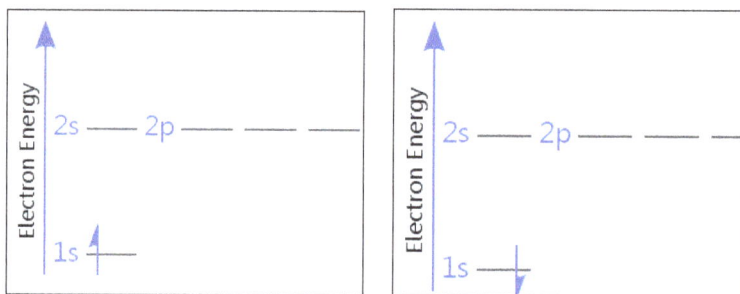

EXCITED STATE

The excited state describes an atom, ion or molecule with an electron in a higher than normal energy level than its ground state.

The length of time a particle spends in the excited state before falling to a lower energy state varies. Short duration excitation usually results in release of a quantum of energy, in the form of a photon or phonon. The return to a lower energy state is called decay. Fluorescence is a fast decay process, while phosphorescence occurs over a much longer time frame. Decay is the inverse process of excitation.

An excited state that lasts a long time is called a metastable state. Examples of metastable states are single oxygen and nuclear isomers.

Sometimes the transition to an excited state enables an atom to participate in a chemical reaction. This is the basis for the field of photochemistry.

Excited State Atom

Eg., consider a carbon atom whose electron configuration is the following.

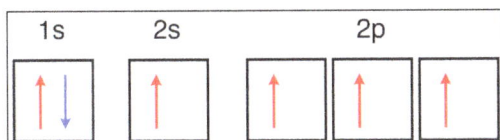

The total energy of the electrons in this carbon atom can be lowered by transfering an electron from a 2P orbital to the 2S orbital. Therefore, this carbon atom is an excited-state carbon atom.

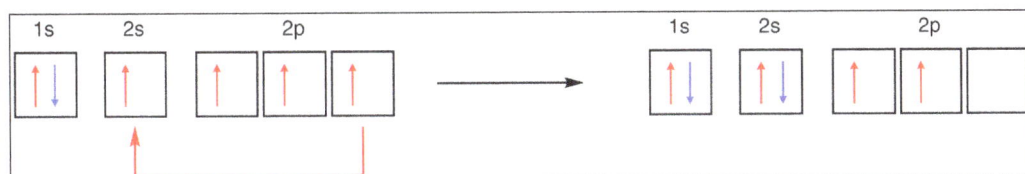

INTERACTION OF AN ATOM WITH RADIATION

In considering the interaction of an atom with radiation, there are three processes to analyze. First, just as a classical oscillating charge will radiate spontaneously, an atom can make a spontaneous transition from an excited state to a state of lower energy, emitting a photon which is the quantum of the electromagnetic field. This process is called "Spontaneous Emission". Second, an atom can absorb a photon from a beam of radiation, making a transition from a state of lower to a state of higher energy. The rate of absorption is proportional to the intensity of the applied field. Finally, atoms can also emit photons under the influence of an applied radiation field. This is called "Stimulated Emission" and it is distinguished from spontaneous emission because the transition rate, like that for absorption, is proportional to the intensity of the applied field. Stimulated emission finds an important practical application in the laser which produces intense beams of coherent radiation. In a rigorous treatment, we would have to start by studying quantum electrodynamics, in which the electromagnetic field is expressed in terms of its quanta- the photons. Each photon corresponding to a field of frequency v carries an amount of energy . Even in comparatively weak fields the photon density can be very high and under these circumstances the number of photons can be treated as a continuous variable and the field can be described classically by the usual Maxwell equations. We shall proceed by using a semi-classical model in which the radiation field is treated classically but the atomic system is described by quantum mechanics. The approximation will also be made that the influence of the atom on the external field can be neglected. Clearly these assumptions do not hold in the case of spontaneous emission, because only one photon is concerned- and one is not a large number! The proper treatment of spontaneous emission is well understood, but is beyond the scope of this discussion. However, in this case, we shall be able to find the transition rate indirectly using a statistical argument due to Einstein.

The classical electromagnetic field is described by electric and magnetic field vectors $\vec{\xi}$ and $\vec{\beta}$, which satisfy Maxwell's equations. We shall express these and other electromagnetic quantities in rationalized MKS units, which form part of the standard "SI" system. The electric field $\vec{\xi}$ and magnetic field $\vec{\beta}$, can be generate from scalar and vector potentials ϕ and $\vec{A}$ by:

$$\vec{\xi}(r,t) = -\nabla\phi(r,t) - \frac{\partial}{\partial t}\vec{A}(r,t)$$

$$\vec{\beta}(\vec{r},t) = \nabla \times \vec{A}(r,t)$$

The potentials are not completely defined by above equations. In particular, $\vec{\xi}$ and $\vec{\beta}$, are unaltered by the substitutions:

$$\vec{A} \rightarrow \vec{A} + \nabla x$$

$$\phi \rightarrow \phi - \frac{\partial x}{\partial t}$$

where x is any scalar field. This property of gauge invariance allows us to impose a further condition on $\vec{A}$, which we shall choose to be:

$$\nabla \cdot \vec{A} = 0$$

When $\vec{A}$ satisfies this condition, we are said to be using the "Coulomb Gauge". From Maxwell's equations we can show that A satisfies the wave equation (as do ϕ, $\vec{\beta}$ and $\vec{\xi}$):

$$\nabla^2\vec{A} - \frac{1}{c^2} + \frac{\partial^2\vec{A}}{\partial t^2} = 0$$

In what follows we shall set the scalar potential $\phi = 0$ since in empty space the most general solution of Maxwell's equations for a radiation field can always be expressed in terms of potentials such that $\nabla \cdot \vec{A} = 0$ and $\phi = 0$. A monochromatic plane wave solution of equations (3) and (4) corresponding to the angular frequency ω (i.e. the frequency $v = \omega/2\pi$) is one that represents a real vector potential $\vec{A}$ as:

$$\vec{A}(\omega;r,t) = 2\vec{A}_0(\omega)\cos(k \cdot r - \omega t + \delta_\omega)$$

$$= \vec{A}_0(\omega)\left[exp[i(k \cdot r - \omega t + \delta_\omega)] + c.c.\right]$$

Here $\vec{A}_0$ is a vector which, describes both the intensity and the polarization of the radiation, $\vec{K}$ is the propagation vector, δ_ω is a real phase and c.c. denotes the complex conjugate. We note that $\nabla \cdot A = 0$ is satisfied if $\vec{k} \cdot \vec{A}_0(\omega) = 0$, so that $\vec{A}_0(\omega)$ is perpendicular to $\vec{K}$ and the wave is transverse. Eq.(4) is satisfied provided that $\omega = kc$, where K is the magnitude of the propagation vector $\vec{K}$. The electric and magnetic fields associated with the vector potential $\vec{A}(\omega;r,t)$, are given respectively from $\vec{\xi}(r,t)$, $\vec{\beta}(r,t)$ equations by:

$$\vec{\xi} = -2\omega\vec{A}_0(\omega)\hat{\varepsilon}\sin(\vec{k} \cdot r - \omega t + \delta_\omega)$$

$$\vec{\beta} = -2\vec{A}_0(\omega)(\vec{k} \times \hat{\varepsilon})\sin(veck \cdot r - \omega + \delta)$$

where we have written $\vec{A}_0(\omega) = \vec{A}_0(\omega)\hat{\varepsilon}$. The direction of the electric field $\vec{\xi}$ is long the real unit vector $\hat{\varepsilon}$, which specifies the polarization of the radiation, and is called "the polarization vector". From $\vec{k} \cdot A_0(\omega) = 0$, we note that $\hat{\varepsilon}$ must lies in a plane perpendicular to the propagation vector $\vec{K}$ and can therefore be specified by giving its components along two linearly independent vectors lying in this plane. We also see from eqs.(6) that both $\vec{\xi}$ and $\vec{\beta}$ are perpendicular to the direction of propagation $\vec{K}$, and to each other.

The expression eqs.(6) describe a "linearly polarized" plane wave, namely a plane wave with its electric field vector $\vec{\xi}$ always in the direction of the polarization vector $\hat{\varepsilon}$. In quantum description of the electromagnetic field the energy in each mode of angular frequency ω , in some region of volume V , is carried by a number $N(\omega)$ of photons, each of energy $\hbar\omega$. The total energy in the mode is therefore given by $N(\omega)\hbar\omega$ and the energy density by $N(\omega)\hbar\omega/V$. In order to relate this quantum description with the classical approach we are using here, we first construct the energy density of the field, which is given by:

$$1/2(\varepsilon_0\vec{\xi}^2 + \vec{\beta}^2/\mu_0) = 4_{\varepsilon 0\omega}{}^2\vec{A_0}^2(\omega)\sin^2(\vec{k} \cdot r - \omega t + \delta_\omega)$$

where ϵ_0 and μ_0 are respectively the permitivity and permeability of free space. The average energy density over a period $2\pi/\omega$ is:

$$\rho(\omega) = 2\varepsilon_0\omega^2\vec{A}_0^2(\omega)$$

Equating this result with the energy density $N(\omega)\hbar\omega/V$, we have:

$$\vec{A}_0^2(\omega) = \frac{\hbar}{2\varepsilon_0\omega V}N(\omega)$$

Similarly, the magnitude of the Poynting Vector $(\vec{\xi} \times \vec{\beta})/\mu_0$ is the rate of energy flow through a unit cross-sectional area normal to the direction of propagation K. Average over a period, this quantity denotes "the intensity of the radiation", which is given by:

$$I(\omega) - 2\varepsilon_0\omega^2 cA_0^2(\omega)\$ = \$\left[\frac{N(\omega)\hbar\omega}{V}\right]c$$

$$= \rho(\omega)c$$

what we have got here, is the intensity of radiation per unit angular frequency.

A general pulse of radiation can by taking $\phi = 0$ and representing $\vec{A}(r,t)$ as a superposition of the plane waves $\vec{A}(\omega;r,t)$. Taking each plane wave component to have the same direction of propagation $\hat{k}$ and adopting a given direction of linear polarization $\hat{\varepsilon}$, so that:

$$\vec{A}_0(\omega) = \vec{A}_0(\omega)\hat{\varepsilon}$$

we write:

$$\vec{A}(r,t) = \int_{\Delta\omega} \vec{A}_0(\omega)\hat{\varepsilon}\left[\exp[i(\vec{k}\cdot r - \omega t + \delta_\omega)] + c.c.\right]$$

We shall be concerned wit the case in which the radiation is nearly monochromatic, so that the amplitude $\vec{A}_0(\omega)$ is peaked about some angular frequency ω_0, differing from zero in a region of width   . In a naturally occurring pulse the radiation arises from many atoms emitting photons independently, which implies that the phases δ_ω are distributed at random, as a function of ω ; in other words, the radiation is "incoherent". It follows that the average energy density in a pulse of the form eq.(11) is:

$$\bar{\rho} = \int_{\Delta\omega} 2\varepsilon_0\omega^2 \vec{A}_0^2(\omega)d(\omega) = \int_{\Delta\omega}\rho(\omega)d(\omega)$$

[This remark does not apply to radiation from lasers, which exhibits a high degree of coherence.]

As seen by comparing with eq. (8) , the contribution from each mode are summed, with no interference term. Similarly the average intensity is:

$$\bar{I} = \int_{\Delta\omega} I(\omega)d\omega$$

where the intensity per unit angular frequency range $I(\omega)$ is given by:

$$I(\omega) = \rho(\omega)c$$

ABSORPTION OF RADIATION, SPONTANEOUS EMISSION AND STIMULATED EMISSION

Every object in the universe is made up of atoms. Atoms are made up of extremely small particles such as electrons, protons, and neutrons. Electrons are the negatively charged particles and protons are the positively charged particles. Neutrons have no charge. Hence, neutrons are referred as neutral particles.

The strong nuclear force between the protons and neutrons makes them stick together to form the nucleus. Neutrons have no charge. so the overall charge of the nucleus is positive because of the protons.

The electrostatic force of attraction between the nucleus and electrons causes electrons to revolve around the nucleus.

The electrons revolving around the nucleus have different energy levels based on the distance from the nucleus.

The electrons revolving very close to the nucleus have lowest energy level whereas the electrons revolving at the farthest distance from nucleus have highest energy level.

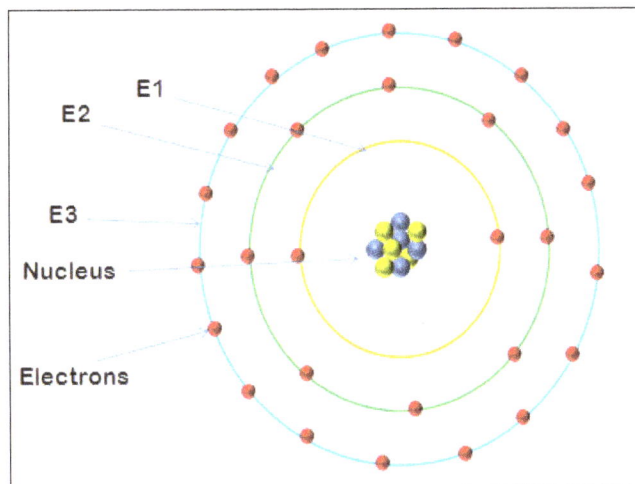

For example, if the lowest energy level is E_1 the next energy level is E_2 and next is E_3, E_4 and so on.

The electrons in the lower energy state (E_1) needs extra energy to jump into next higher energy state (E_2). This energy can be supplied in the form of the electric field, heat or light.

When the electrons in the lower energy state (E_1) gains sufficient energy from photons, they jump into next higher energy state (E_2).

The electrons in the higher energy state do not stay for long period. After a short period, they again fall back to the lower energy level by losing their energy. The electrons in the higher energy level or higher energy state lose energy in the form of light before they fall back to the lower energy state.

The electrons in the higher energy state are known as excited electrons whereas the electrons in the lower energy state are known as ground electrons.

In lasers, the way light or photons interact with atoms plays an important role in its operation. The photons interact in three ways with the atoms:

- Absorption of radiation or light.

- Spontaneous emission.

- Stimulated emission.

Absorption of Radiation or Light

The process of absorbing energy from photons is called absorption of radiation.

It is well known that there are different energy levels in an atom. The electrons that are very close to the nucleus have lowest energy level. These electrons are also known as ground state electrons.

Let us consider that the energy level of ground state electrons or lower energy state electrons is E_1 and the next higher energy level or higher energy state is E_2.

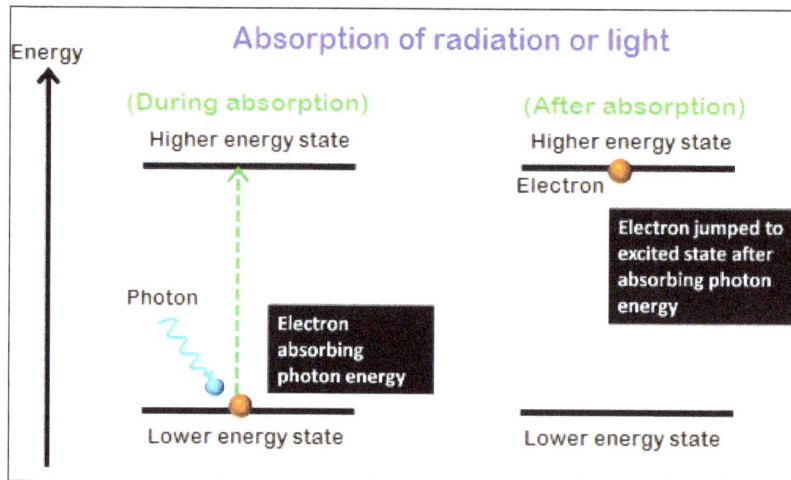

When ground state electrons or lower energy state electrons (E_1) absorbs sufficient energy from photons, they jump into the next higher energy level or higher energy state (E_2). In other words, when the ground state electrons absorb energy which is equal to the energy difference between the two energy states ($E_2 - E_1$), the electrons jumps from ground state (E_1) to the excited state or higher energy level (E_2). The electrons in the higher energy level are called excited electrons.

The light or photons energy applied to excite the electrons can be mathematically written as:

$$hv = E_2 - E_1$$

Where h = Planck's constant.

V = Frequency of photon.

E_1 = Lower energy level electrons or ground state electrons.

E_2 = Higher energy level electrons or excited state electrons.

Absorption occurs only if the energy of photon exactly matches the difference in energy between the two electron shells or orbits.

Spontaneous Emission

The process by which excited electrons emit photons while falling to the ground level or lower energy level is called spontaneous emission.

Electrons in the atom absorb energy from various sources such as heat, electric field, or light. When the electrons in the ground state or lower energy state (E_1) absorb sufficient energy from photons, they jump to the excited state or next higher energy state (E_2).

The electrons in the excited state do not stay for a long period because the lifetime of electrons in the higher energy state or excited state is very small, of the order of 10-8 sec. Hence, after a short period, they fall back to the ground state by releasing energy in the form of photons or light.

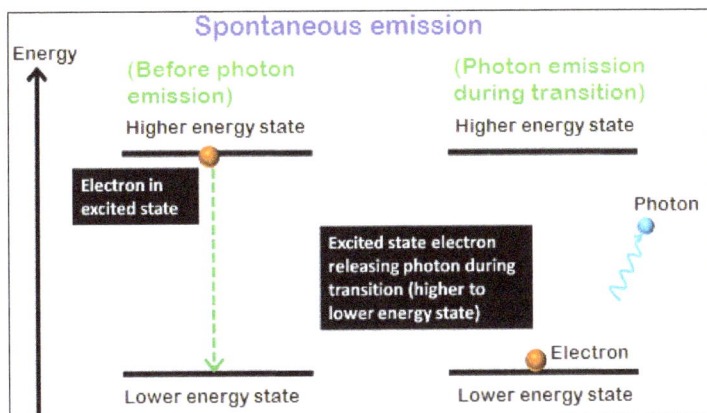

The energy of the emitted photon is directed proportional to the energy gap of the material. The materials with large energy gap will emit high-energy photons or high-intensity light whereas the materials with small energy gap will emit low energy photons or low-intensity light.

The energy of released photon is equal to the difference in energies between the two electron shells or orbits.

The energy of the excited electrons can also be released in other forms such as heat. If the excited state electrons release energy in the form of photons or light while falling to the ground state, the process is called spontaneous emission.

In spontaneous emission, the electrons changing from one state (higher energy state) to another state (lower energy state) occurs naturally. So the photon emission also occurs naturally or spontaneously.

The photons emitted due to spontaneous emission do not flow exactly in the same direction of incident photons. They flow in the random direction.

Stimulated Emission

The process by which electrons in the excited state are stimulated to emit photons while falling to the ground state or lower energy state is called stimulated emission.

Unlike the spontaneous emission, in this process, the light energy or photon energy is supplied to the excited electrons instead of supplying energy to the ground state electrons.

The stimulated emission is not a natural process it is an artificial process. In stimulated emission, the electrons in the excited state need not wait for natural spontaneous emission to occur. An alternative method is used to stimulate excited electron to emit photons and fall back to ground state.

The incident photon stimulates or forces the excited electron to emit a photon and fall into a lower state or ground state.

The energy of a stimulating or incident photon must be equal to the energy difference between the two electron shells.

In stimulated emission process, each incident photon generates two photons.

The photons emitted in the stimulated emission process will travel in the same direction of the incident photon.

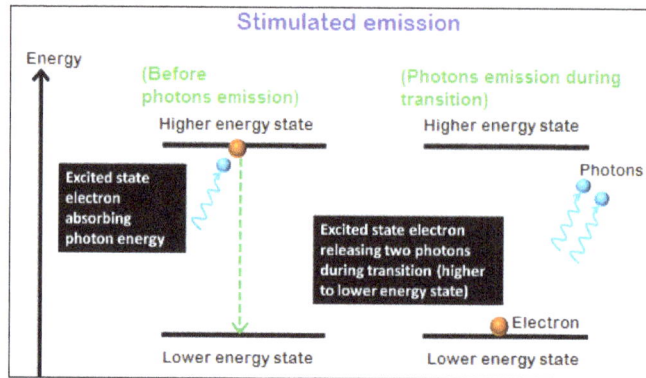

In this process, the excited electron releases an additional photon of same energy (same frequency, same phase, and in the same direction) while falling into the lower energy state. Thus, two photons of same energy are released while electrons falling into the ground state.

Many ways exist to produce light, but the stimulated emission is the only method known to produce coherent light (beam of photons with the same frequency).

All the photons in the stimulated emission have the same frequency and travel in the same direction.

EMISSION AND ABSORPTION SPECTRA

Emission Spectra

The electrons surrounding the atomic nucleus are arranged in a series of levels of increasing energy. Each element has a unique number of electrons in a unique configuration therefore each element has its own distinct set of energy levels. This arrangement of energy levels serves as the atom's unique fingerprint.

In the early 1900s, scientists found that a liquid or solid heated to high temperatures would give off a broad range of colours of light. However, a gas heated to similar temperatures would emit light only at certain specific wavelengths (colours). The reason for this observation was not understood at the time.

Scientists studied this effect using a discharge tube.

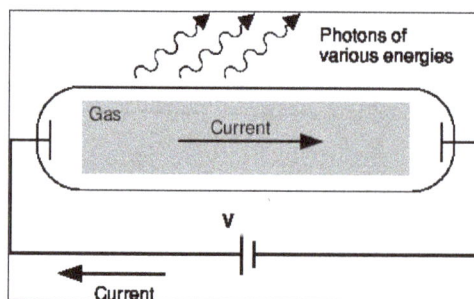

Diagram of a discharge tube. The tube is filled with a gas. When a high enough voltage is applied across the tube, the gas ionises and acts like a conductor, allowing a current to flow through the circuit. The current excites the atoms of the ionised gas. When the atoms fall back to their ground state, they emit photons to carry off the excess energy.

A discharge tube is a gas-filled, glass tube with a metal plate at both ends. If a large enough voltage difference is applied between the two metal plates, the gas atoms inside the tube will absorb enough energy to make some of their electrons come off, i.e. the gas atoms are ionised. These electrons start moving through the gas and create a current, which raises some electrons in other atoms to higher energy levels. Then as the electrons in the atoms fall back down, they emit electromagnetic radiation (light). The amount of light emitted at different wavelengths, called the emission spectrum, is shown for a discharge tube filled with hydrogen gas in figure below. Only certain wavelengths (i.e., colours) of light are seen, as shown by the lines in the picture.

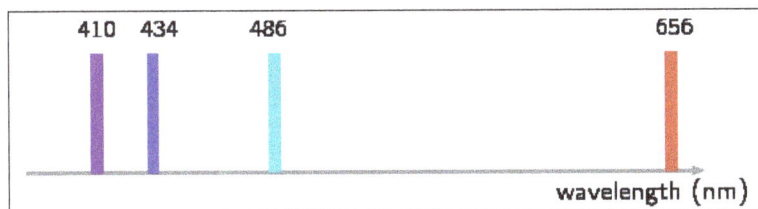

Diagram of the emission spectrum of hydrogen in the visible spectrum. Four lines are visible, and are labelled with their wavelengths. The three lines in the 400–500 nm range are in the blue part of the spectrum, while the higher line (656 nm) is in the red/orange part.

Eventually, scientists realised that these lines come from photons of a specific energy, emitted by electrons making transitions between specific energy levels of the atom. Figure shows an example of this happening. When an electron in an atom falls from a higher energy level to a lower energy level, it emits a photon to carry off the extra energy. This photon's energy is equal to the energy difference between the two energy levels (ΔE).

$$\Delta E electron = E_f - E_i$$

The frequency of a photon is related to its energy through the equation $E = hf$. Since a specific photon frequency (or wavelength) gives us a specific colour, we can see how each coloured line is associated with a specific transition.

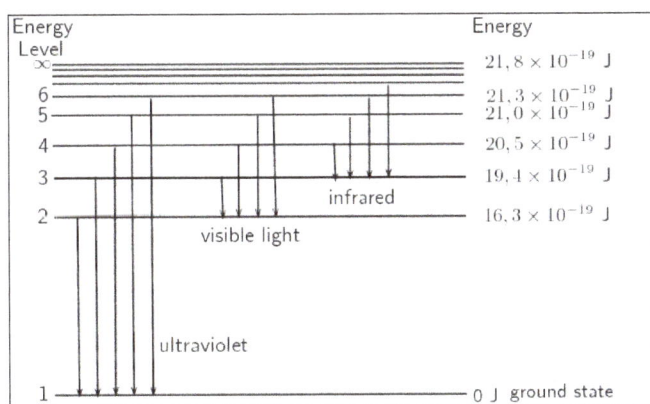

The arrows show the electron transitions from higher energy levels to lower energy levels. The energies of the emitted photons are the same as the energy difference between two energy levels. You can think of absorption as the opposite process. The arrows would point upwards and the electrons would jump up to higher levels when they absorb a photon of the right energy. The second representation shows the wavelengths of the light that is emitted for the the various transitions. The transistions are grouped into a series based on the lowest level involved in the transition.

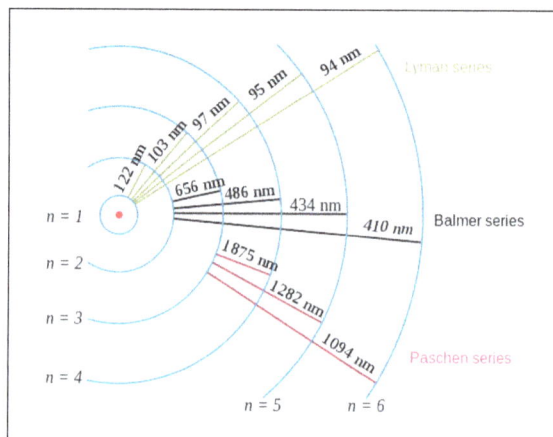

In the first diagram are shown some of the electron energy levels for the hydrogen atom.

Visible light is not the only kind of electromagnetic radiation emitted. More energetic or less energetic transitions can produce ultraviolet or infrared radiation. However, because each atom has its own distinct set of energy levels (its fingerprint), each atom has its own distinct emission spectrum.

Absorption Spectra

Atoms do not only emit photons; they also absorb photons. If a photon hits an atom and the energy of the photon is the same as the gap between two electron energy levels in the atom, then the electron in the lower energy level can absorb the photon and jump up to the higher energy level. If the photon energy does not correspond to the difference between two energy levels then the photon will not be absorbed (it can still be scattered).

Using this effect, if we have a source of photons of various energies we can obtain the absorption spectra for different materials. To get an absorption spectrum, just shine white light on a sample of the material that you are interested in. White light is made up of all the different wavelengths of visible light put together. In the absorption spectrum there will be gaps. The gaps correspond to energies (wavelengths) for which there is a corresponding difference in energy levels for the particular element.

The absorbed photons show up as black lines because the photons of these wavelengths have been absorbed and do not show up. Because of this, the absorption spectrum is the exact inverse of the emission spectrum. Look at the two figures below. In figure you can see the line emission spectrum of hydrogen. Figure shows the absorption spectrum. It is the exact opposite of the emission spectrum! Both emission and absorption techniques can be used to get the same information about the energy levels of an atom.

Emission spectrum of Hydrogen.

Absorption spectrum of Hydrogen.

The dark lines correspond to the frequencies of light that have been absorbed by the gas. As the photons of light are absorbed by electrons, the electrons move into higher energy levels. This is the opposite process of emission.

The dark lines, absorption lines, correspond to the frequencies of the emission spectrum of the same element. The amount of energy absorbed by the electron to move into a higher level is the same as the amount of energy released when returning to the original energy level.

Applications of Emission and Absorption Spectra

Spectroscopy is a tool widely used in astronomy to learn different things about astronomical objects.

Identifying Elements in Astronomical Objects using their Spectra

Measuring the spectrum of light from a star can tell astronomers what the star is made of. Since each element emits or absorbs light only at particular wavelengths, astronomers can identify what elements are in the stars from the lines in their spectra. From studying the spectra of many stars we know that there are many different types of stars which contain different elements and in different amounts.

Determining Velocities of Galaxies using Spectroscopy

The same thing happens to electromagnetic radiation (light). If the object emitting the light is moving towards us, then the wavelength of the light appears shorter (called blueshifted). If the object is moving away from us, then the wavelength of its light appears stretched out (called redshifted).

The Doppler effect affects the spectra of objects in space depending on their motion relative to us on the earth. For example, the light from a distant galaxy that is moving away from us at some velocity will appear redshifted. This means that the emission and absorption lines in the galaxy's spectrum will be shifted to a longer wavelength (lower frequency). Knowing where each line in the spectrum would normally be if the galaxy was not moving and comparing it to the redshifted position, allows astronomers to precisely measure the velocity of the galaxy relative to the Earth.

Global Warming and Greenhouse Gases

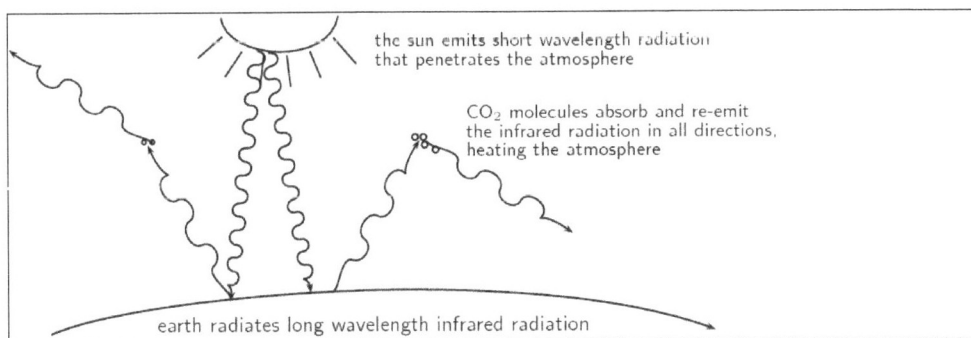

the sun emits short wavelength radiation that penetrates the atmosphere

CO_2 molecules absorb and re-emit the infrared radiation in all directions, heating the atmosphere

earth radiates long wavelength infrared radiation

The sun emits radiation (light) over a range of wavelengths that are mainly in the visible part of the spectrum. Radiation at these wavelengths passes through the gases of the atmosphere to warm the land and the oceans below. The warm earth then radiates this heat at longer infrared wavelengths. Carbon dioxide (one of the main greenhouse gases) in the atmosphere has energy levels that correspond to the infrared wavelengths that allow it to absorb the infrared radiation. It then also emits at infrared wavelengths in all directions. This effect stops a large amount of the infrared radiation from getting out of the atmosphere, which causes the atmosphere and the earth to heat up. More radiation is coming in than is getting back out.

So increasing the amount of greenhouse gases in the atmosphere increases the amount of trapped infrared radiation and therefore the overall temperature of the earth. The earth is a very sensitive and complicated system upon which life depends and changing the delicate balances of temperature and atmospheric gas content may have disastrous consequences if we are not careful.

EXCITATION

Excitation, in physics is the addition of a discrete amount of energy (called excitation energy) to a system—such as an atomic nucleus, an atom, or a molecule—that results in its alteration, ordinarily from the condition of lowest energy (ground state) to one of higher energy (excited state).

In nuclear, atomic, and molecular systems, the excited states are not continuously distributed but have only certain discrete energy values. Thus, external energy (excitation energy) can be absorbed only in correspondingly discrete amounts.

Thus, in a hydrogen atom (composed of an orbiting electron bound to a nucleus of one proton), an excitation energy of 10.2 electron volts is required to promote the electron from its ground state to the first excited state. A different excitation energy (12.1 electron volts) is needed to raise the electron from its ground state to the second excited state.

Similarly, the protons and neutrons in atomic nuclei constitute a system that can be raised to discrete higher energy levels by supplying appropriate excitation energies. Nuclear excitation energies are roughly 1,000,000 times greater than atomic excitation energies. For the nucleus of lead-206, as an example, the excitation energy of the first excited state is 0.80 million electron volts and of the second excited state 1.18 million electron volts.

The excitation energy stored in excited atoms and nuclei is radiated usually as visible light from atoms and as gamma radiation from nuclei as they return to their ground states. This energy can also be lost by collision.

The process of excitation is one of the major means by which matter absorbs pulses of electromagnetic energy (photons), such as light, and by which it is heated or ionized by the impact of charged particles, such as electrons and alpha particles. In atoms, the excitation energy is absorbed by the orbiting electrons that are raised to higher distinct energy levels. In atomic nuclei, the energy is absorbed by protons and neutrons that are transferred to excited states. In a molecule, the energy is absorbed not only by the electrons, which are excited to higher energy levels, but also by the whole molecule, which is excited to discrete modes of vibration and rotation.

INTERACTION OF ATOMS WITH A CLASSICAL LIGHT FIELD

The Electron Wavefunction and the Two-level Atom

The wavefunction of an electron $\Psi(x)$ can be decomposed with a complete set of eigenfunctions $\psi_j(x)$ which obey the Schroedinger equation:

$$H_0\psi_j(x) = \left(-\frac{\hbar^2}{2m}\nabla^2 + V\right)\psi_j(x) = E_j\psi_j(x)$$

In analogy to the quantization of the light field one can write:

$$\Psi(x) = \sum_j b_j^+\psi_j(x)$$

with the fermionic creation operator b_j^+.

The anti-commutation relation of the fermionic creation and annihilation operator are:

$$\{b_i, b_j\} = \{b_i^+, b_j^+\} = 0$$

$$\{b_j, b_j^+\} = 1$$

An arbitrary state can thus be constructed by applying b_j^+ operators to the vacuum:

$$\left|\{j\}\right\rangle = b_{j1}^+ b_{j2}^+ .... b_{jn}^+ |0\rangle$$

Due to the fermionic nature:

$$\left(b_j^+\right)^2|0\rangle = 0 \text{ or more general } \left(b_j^+\right)^2|\varphi\rangle = 0$$

The expectation value for the atomic Hamiltonian H_0

$$H_0 = \sum_j b_j^+ b_j E_j$$

Is,

$$\langle\psi|H_0|\psi\rangle = \sum_j E_j$$

Since a lot of problems in quantum optics deal with the simplified case of two-level atoms it is convenient to limit the atomic Hilbert space to two dimensions and to introduce the Pauli spin operators σj $\in H \oplus 2$ (similar as in a single spin system):

$$\sigma_x = \begin{pmatrix} 0 & 1 \\ 1 & 0 \end{pmatrix}; \ \sigma_y = \begin{pmatrix} 0 & -i \\ i & 0 \end{pmatrix}; \ \sigma_z = \begin{pmatrix} 1 & 0 \\ 0 & -1 \end{pmatrix}$$

Together with the raising and lowering operators,

$$\sigma^+ = \frac{1}{2}\left(\sigma_x + i\sigma_y\right); \; \sigma^- = \frac{1}{2}\left(\sigma_x - i\sigma_y\right)$$

The latter operators have the following properties:

$$\left[\sigma^+, \sigma^-\right] = 2\sigma_z; \left[\sigma^\pm, \sigma_z\right] = \mp\sigma^\pm; \left\{\sigma^+, \sigma^-\right\} = 1$$

Bloch Representation

If we assume a two-level system of two atomic states $|1\rangle = \begin{pmatrix} 1 \\ 0 \end{pmatrix}$ and $|2\rangle = \begin{pmatrix} 0 \\ 1 \end{pmatrix}$ then the following correspondence holds:

Pseudo-spin operators	electron operators:	
σ^+	$b_1^+ b_2$	$|1\rangle\langle 2|$
σ^-	$b_2^+ b_1$	$|2\rangle\langle 1|$

Any state of the two-level atom can be written as:

$$|\psi\rangle = c_1|1\rangle + c_2|2\rangle \; \text{ with } \; |c_1|^2 + |c_2|^2 = 1$$

More generally (non-pure states) one has to write down the density operator ρ:

$$\rho(A) = \rho_{11}|1\rangle + \rho_{22}|2\rangle\langle 2| + \rho_{12}|1\rangle\langle 2| + \rho_{21}|2\rangle\langle 1|$$

Where:

$$\rho_{ij} = \langle c_i c_j^*\rangle \; i, j = 1, 2$$

ρ has a representation in terms of a two-dimensional Hermitian covariant matrix. The Bloch-representation has a very intuitive geometrical representation of the state.

Definition of the Bloch-vector $\vec{r}$:

$$r_1 = 2\,\mathrm{Re}\left(\rho_{12}\right)$$
$$r_2 = 2\,\mathrm{Im}\left(\rho_{12}\right)$$
$$r_3 = \rho_{22} - \rho_{11}$$

Therefore:

$$|1\rangle \triangleq (0,0,-1)$$
$$|2\rangle \triangleq (0,0,1)$$

The Bloch-vector for a pure state lies on a sphere of radius $|r| = 1$.

Generally, it follows:

$$r_1^2 + r_2^2 + r_3^2 = 4|\rho_{12}|^2 + |\rho_{22} - \rho_{11}|^2$$
$$= 1 - 4\left(\rho_{22}\rho_{11} - |\rho_{12}|^2\right)$$

From the Cauchy-Schwartz inequality on finds:

$$\rho_{22}\rho_{11} - |\rho_{12}|^2 = \left\langle |c_2|^2 \right\rangle \left\langle |c_1|^2 \right\rangle \left\langle |c_1|^2 \right\rangle - \left|\left\langle c_1 c_2^* \right\rangle\right|^2 \geq 0$$

and thus:

$$r_1^2 + r_2^2 + r_3^2 \leq 1$$

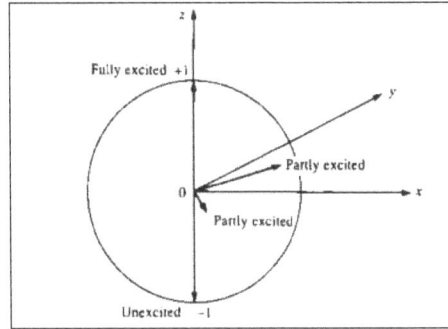

Bloch representation of a state of a two-level atom.

Interaction of a Two-level Atom with a Classical Field

The interaction of a classical field E(t) with an atom can be described via the dipole interaction:

$$H_I = -\vec{\mu}(t).\vec{E}(t)$$

The time evolution of the density matrix $\rho(t)$ describing the state of the atom (Hamiltonian $H_A = \frac{1}{2}\hbar\omega_0\sigma_z$ follows from the Schroedinger equation with the Hamiltonian $H = H_A + H_I$:

$$\frac{\partial p(t)}{\partial t} = \frac{1}{i\hbar}\left[H_A + H_I, \rho(t)\right]$$

The general form of this equation of motion is:

$$\dot{\rho}_{11} = \frac{1}{i\hbar}\left[\left\langle 1|H_I|2 \right\rangle \rho_{21} - c.c\right]$$

$$\dot{\rho}_{22} = \frac{1}{i\hbar}\left[\left\langle 1|H_I|2 \right\rangle \rho_{21} - c.c\right]$$

$$\dot{\rho}_{12} = \frac{1}{i\hbar}\left[-\hbar\omega_0\rho_{12} + \left\langle 1|H_I|2 \right\rangle(\rho_{22} - \rho_{11})\right]$$

$$\dot{\rho}_{21} = \frac{1}{i\hbar}\left[\hbar\omega_0\rho_{21} + \left\langle 2|H_I|1 \right\rangle(\rho_{11} - \rho_{22})\right]$$

Obviously $(\dot{\rho}_{11} + \dot{\rho}_{22}) = 0$.

Remark: The link to classical or semiclassical physics is via the polarisation,

$$P = \langle 1|H_I|2 \rangle \rho_{12} + c.c.$$

These equations of motions can be expressed by the Bloch vector and are called Bloch equations:

$$\dot{r}_1 = \frac{1}{\hbar} 2 \operatorname{Im}\left[\langle 1|H_I|2 \rangle\right] r_3 - \omega_0 r_2$$

$$\dot{r}_2 = -\frac{1}{\hbar} 2 \operatorname{Re}\left[\langle 1|H_I|2 \rangle 2\right] r_3 + \omega_0 r_1$$

$$\dot{r}_3 = -\frac{2}{\hbar} \operatorname{Im}\left[\langle 1|H_I|2 \rangle\right] r_1 + \frac{2}{\hbar} \operatorname{Re}\left[\langle 1|H_I|2 \rangle\right] r_2$$

Obviously $d/dt\left(r_1^2 + r_2^2 + r_3^2\right) = 0$!

The motion of the Bloch vector can be described as a (complicated) precession around a vector $Q(t)$:

$$\frac{d}{dt}\vec{r} = Q \times \vec{r}$$

with,

$$Q = \begin{pmatrix} \dfrac{2}{\hbar} \operatorname{Re}\langle 1|H_I|2 \rangle \\ \dfrac{2}{\hbar} \operatorname{Im}\langle 1|H_I|2 \rangle \\ \omega_0 \end{pmatrix}$$

If the interaction with a classical single-mode field $E(t) = \hat{\varepsilon} E_0(t) \exp(-i\omega_1 t) + c.c.$ is evaluated then the term h1| H_I |2i becomes:

$$\langle 1|H_I|2 \rangle = -\vec{\mu}_{12}\vec{E}(t) = -\langle 1|\vec{\mu}|2 \rangle \vec{E}(t)$$

The fast rotation of the Bloch-vector around the z-axis at the optical frequency $\omega 0$ can be eliminated by transforming into a rotating frame:

$$\vec{r}' = \Theta \cdot \vec{r}$$

With,

$$\Theta = \begin{pmatrix} \cos \omega_1 t & \sin \omega_1 t & 0 \\ -\sin \omega_1 t & \cos \omega_1 t & 0 \\ 0 & 0 & 1 \end{pmatrix}$$

This leads to the Bloch equations in the rotating frame:

$$\dot{r_1'} = (\omega_1 - \omega_0) r_2'$$
$$\dot{r_2'} = (\omega_0 - \omega_1) r_1' + \Omega r_3'$$
$$\dot{r_3'} = -\Omega r_2'$$

with the Rabi frequency Ω:

$$\Omega = 2\vec{\mu}\, 12^{\hat{\varepsilon}} \left| E_0(t) \right| / \hbar$$

One can also write:

$$\vec{r}' = Q' \times \vec{r}' \text{ with } Q' = \begin{pmatrix} -\Omega \\ 0 \\ \omega_0 - \omega_1 \end{pmatrix}$$

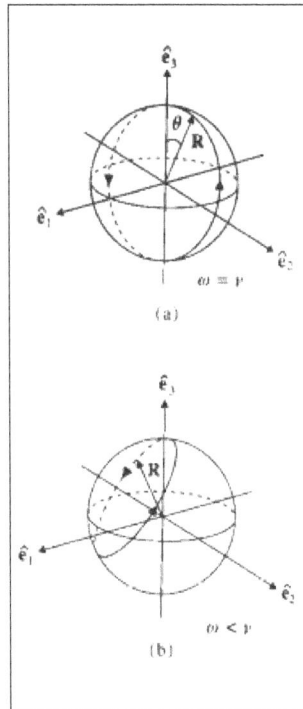

Precession of Bloch vector vor $\delta = 0$ (a) and $\delta\, 6= 0$ (b).

Ramsey Fringes

If the field is in resonance with the atomic transition $(\omega 1 - \omega 0) = 0$ then it is:

$$r_1'(t) = 0$$
$$r_2'(t) = 0 - \sin \Omega t$$
$$r_3'(t) = \cos \Omega t$$

A pulse which is applied to the atom initially in the ground state ($r = (0, 0, 1)$) which has the pulse area $\Omega t = \pi$, a so-called π-pulse, flips the atomic state to the excited state, whereas a pulse with area $\Omega t = \pi/2$, a $\pi/2$-pulse, creates a coherent superposition of upper and lower atomic state of equal weight:

$$\Omega t = \pi \qquad \pi-\text{pulse} \qquad |2\rangle \rightarrow |1\rangle$$
$$\Omega t = \pi/2 \qquad \pi/2-\text{pulse} \qquad |2\rangle \rightarrow \left(|2\rangle + |1\rangle\right)/\sqrt{2}$$

A small detuning $\delta = \omega_1 - \omega_0$ leads to a rotation of the Bloch vector in the x-y-plane if there is a non-zero component of r_1 or r_2.

A method to exploit this effect in order to perform precise measurements of a frequency ω was proposed by Ramsey, who was awarded the Nobel prize for this idea in 1989:

- First a $\pi/2$-pulse is applied to an atom, which is initially in the ground state. This flips the Bloch vector into the x-y-plane.

- If there is no detuning (e.-mag. field in exact resonance with the atomic transition) then a second $\pi/2$-pulse after some time T flips the Bloch vector exactly to the excited state, which can then be detected.

- If, however, there is some detuning then the Bloch vector rotates in the x-yplane by an angle $\delta \cdot T$. A second $\pi/2$-pulse would then usually not tilt the Bloch vector exactly to the excited state (in the extreme case the Bloch vector may even be tilted back to the ground state).

This method can be used to compare the frequency of a field to an atomic transition and is called Ramsey-method. By changing T or δ the probability to detect the atom in the excited state oscillates. These oscillations are also called Ramsey fringes.

In a Ramsey interferometer the two pulses have to be separated in time as far as possible to obtain highest sensitivity. The sensitivity is not limited by the time-of-flight of the atom through a single interaction zone in the experiment. Modern atomic clocks (e.g. Cs clocks) use the Ramsey method to stabilze an RFfield to a narrow atomic transition.

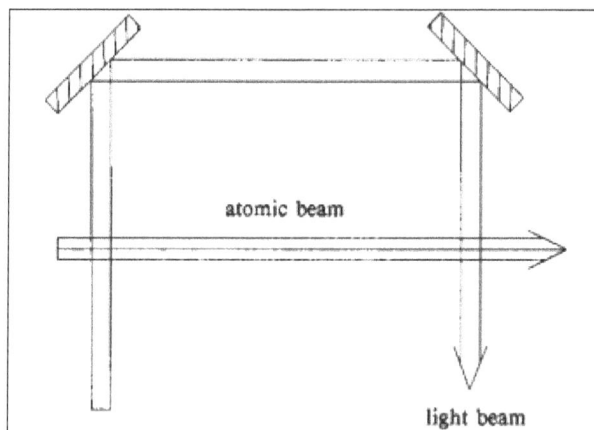

Principle setup for a Ramsey measurement.

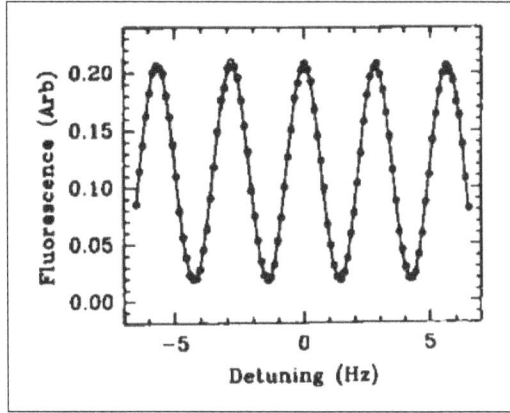

Ramsey fringes.

Rabi-oscillation for Pure States

Rabi-Oscillations are the typical dynamics for the coherent interaction of two-level atoms with light. In the following some aspects are described more explicitly for the case of pure states.

A general state of a two-level atom is of the form:

$$|\psi\rangle = c_e|e\rangle + c_g|g\rangle$$

We are interested in the time evolution of the coefficienty $c_e(t)$ and $c_g(t)$.

Following equation for the density matrix it is possible to write:

$$\rho = p_{ee}|e\rangle\langle e| + p_{gg}|g\rangle\langle g| + \rho_{eg}|e\rangle\langle g| + \rho_{ge}|g\rangle\langle e|$$

Where:

$$\rho_{ij} = c_i c_j^* \quad i,j = e,g$$

where we replaced the previously used subscript 1 and 2 by e and g, respectively, for clarity.

Resonant Interaction

In the special case of exact resonance between the frequency of the light and the atomic transition frequency, i.e. $\delta = \omega_1 - \omega_0 = 0$, it is straightforward to derive an exact solution from the equation of motion derived above:

$$c_e(t) = c_e(0)\cos\left(\frac{1}{2}\Omega t\right) - ic_g(0)\sin\left(\frac{1}{2}\Omega t\right)$$

$$c_g(t) = c_g(0)\cos\left(\frac{1}{2}\Omega t\right) - ic_e(0)\sin\left(\frac{1}{2}\Omega t\right)$$

with the Rabi-frequency $\Omega = 2 < e|H_I|g>/\hbar$.

With the special initial condition $c_e(0) = 0$, $c_g(0) = 1$ one finds:

$$c_e(t) = -\sin\left(\frac{1}{2}\Omega t\right)$$

$$c_g(t) = \cos\left(\frac{1}{2}\Omega t\right)$$

and for the probability $P_g(t)$ *and* $P_e(t)$ to find the atom in the ground and excited state, respectively:

$$P_e(t) = |c_e(t)|^2 = \frac{1}{2}(1 - \cos\Omega t)$$

$$P_g(t) = |c_g(t)|^2 = \frac{1}{2}(1 + \cos\Omega t)$$

Figure plots the time evolution which corresponds to the pictorial dynamics of the Bloch vector as described above.

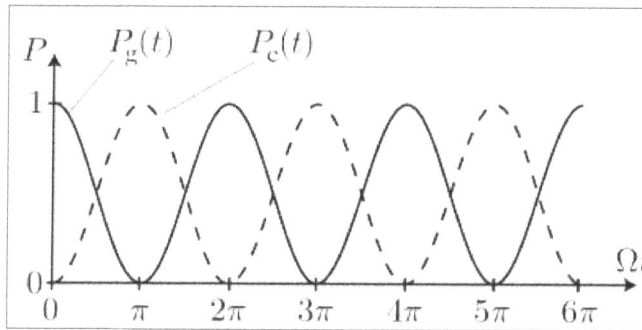

Time evolution of the probability $P_g(t)$ *and* $P_e(t)$ to find the atom in the ground (solid) and excited (dashed) state, respectively.

Phenomenological Treatment of Damping

Generalized Bloch-equation

How can we describe an incoherent evolution, e.g., damping?

The starting points are the Bloch-equation as provided in equation:

$$\dot{\rho}_{11} = \frac{1}{i\hbar}\left[\langle 1|H_I|2\rangle\rho_{21} - c.c\right]$$

$$\dot{\rho}_{22} = \frac{1}{i\hbar}\left[\langle 1|H_I|2\rangle\rho_{21} - c.c\right]$$

$$\dot{\rho}_{12} = \frac{1}{i\hbar}\left[-\hbar\omega_0\rho_{12} + \langle 1|H_I|2\rangle(\rho_{22} - \rho_{11})\right]$$

$$\dot{\rho}_{21} = \frac{1}{i\hbar}\left[\hbar\omega_0\rho_{21} + \langle 2|H_I|1\rangle(\rho_{11} - \rho_{22})\right]$$

We will now introduce damping terms in a phenomenological way:

- Γ as a damping rate of the population (diagonal elements of the density matrix).

- $\gamma \perp$ as a damping rate of the coherence (off-diagonal elements of the density matrix).

The rate Γ also damps the off-diagonal elements. In order to include more general cases we introduce an additional rate γ_c which describes pure dephasing (e.g. due to collisions of two-level atoms in a gas). Therefore:

$$\gamma \perp = \Gamma/2 + \gamma_c$$

With the damping rates we can formulate the generalized version of the optical Bloch equation including damping (in the interaction picture):

$$\dot{\rho}_{ee} = i\frac{\Omega}{2}\left(\rho_{eg} - \rho_{ge}\right) - \Gamma\rho_{ee}$$

$$\dot{\rho}_{gg} = -i\frac{\Omega}{2}\left(\rho_{eg} - \rho_{ge}\right) + \Gamma\rho_{ee}$$

$$\dot{\rho}_{ge} = -\left(\gamma \perp + i\delta\right)\rho_{ge} - i\frac{\Omega}{2}\left(\rho_{ee} - \rho_{gg}\right)$$

$$\dot{\rho}_{eg} = -\left(\gamma \perp - i\delta\right)\rho_{eg} + i\frac{\Omega}{2}\left(\rho ee - \rho_{gg}\right)$$

where we have again introduced the Rabi-frequency $\Omega = 2\langle 1|H_I|2\rangle / \hbar$. In a similar way the equations of motion for the Bloch-vector in the interaction picture can be generalized as:

$$\dot{r_1'} = \delta r_2' - \gamma_\perp r_1'$$
$$\dot{r_2'} = -\delta r_1' + \Omega r_3' - \gamma_\perp r_2'$$
$$\dot{r_3'} = -\Omega r_2' - \Gamma\left(r_3' + 1\right)$$

with $\delta = \omega_1 - \omega_0$

Steady-state-solution of the Bloch-equation with Damping

It is straightforward to derive the steady-state solution of the Bloch-equation by setting the time derivatives to zero.

As a result for the population of excited state ρ_{ee} we find:

$$\rho_{ee} = \frac{\Omega^2}{2\Gamma\gamma_\perp} \frac{1 + \dfrac{i\delta}{\gamma_\perp}}{1 + \dfrac{\delta^2}{\gamma_\perp^2} + \dfrac{\Omega^2}{\Gamma\gamma_\perp}}$$

And for the steady state coherence ρ_{eg}:

$$\rho_{eg} = -i \frac{\Omega}{2\gamma_{\perp}} \frac{1 + \dfrac{i\delta}{\gamma_{\perp}}}{1 + \dfrac{\delta^2}{\gamma_{\perp}^2} + \dfrac{\Omega^2}{\Gamma\gamma_{\perp}}}$$

Often the so-called saturation parameter S is introduced to simplify the notation.

$$S = \frac{\Omega^2 / \Gamma\gamma_{\perp}}{1 + \delta^2 / \gamma_{\perp}^2}$$

With this parameter the expression for the steady-state values of ρ_{ee} and $|\rho_{eg}|^2$ are:

$$\rho_{ee} = \frac{S/2}{1+S}$$

$$|\rho_{eg}|^2 = \frac{\Gamma}{4\gamma_{\perp}} \frac{S}{(1+S)^2}$$

Finally, we give the analytic form for the time evolution of the excited state probability $\rho_{ee}(t)$ for the special case $\delta = 0$, $\gamma_{\perp} = \Gamma/2$ and the atom initially in the ground state.

$$\rho_{ee}(t) = -1 + \frac{\Omega^2}{\Omega^2 + \Gamma^2/2}\left[1 - e^{-(3\Gamma/4)t}\left(\cos\Omega_r t + \frac{\Omega^2 - \Gamma^2/4}{\Gamma\Omega_r}\sin\Omega_r t\right)\right]$$

Where, $\Omega r = \sqrt{\Omega^2 + (\Gamma/4)^2}$.

Note that after some initial oscillation the atomic inversion will relax to a value below 0, i.e. the population in the upper state is always less than the population in the lower state. The following figure plots time evolutions for different values of the damping Γ:

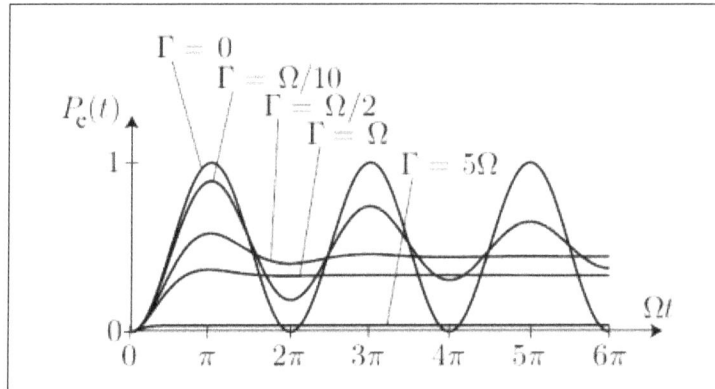

Time evolution of the probability $\rho_{ee}(t)$ to find the atom
in the excited state for different damping rates Γ.

References

- Ground-state, definition: chemicool.com, Retrieved 13 July, 2019

- Definition-of-excited-state: thoughtco.com, Retrieved 1 August, 2019

- Node2, laserandlight, nazila, amos: uib.no, Retrieved 19 May, 2019

- Absorption-of-radiation-spontaneous-emission-and-stimulated-emission: physics-and-radio-electronics.com, Retrieved 28 February, 2019

- Excited-state-atom, alphabetical: ochempal.org, Retrieved 18 April, 2019

- 12-optical-phenomena-and-properties-of-matter, optical-phenomena-and-properties-of-matter, grade-12, read: siyavula.com, Retrieved 2 July, 2019

- Excitation, science: britannica.com, Retrieved 21 April, 2019

CHAPTER 4

Bohr Model of the Atom

Bohr model represents the structure of atom and its constituents. It depicts nucleus at the center and electrons revolve around it. This chapter discusses in detail the theories and methodologies related to Bohr model including Bohr model of atom and hydrogen, hydrogen spectrum, Franck-Hertz experiment, etc.

The Bohr Model has an atom consisting of a small, positively-charged nucleus orbited by negatively-charged electrons. Here's a closer look at the Bohr Model, which is sometimes called the Rutherford-Bohr Model.

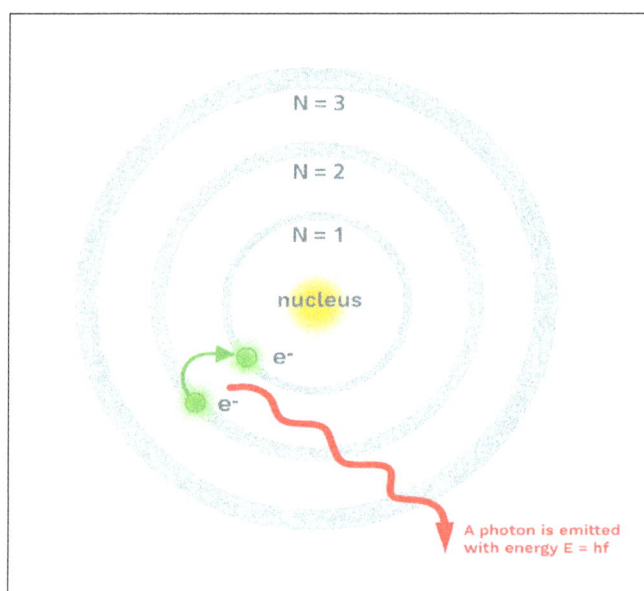

Niels Bohr proposed the Bohr Model of the Atom in 1915. Because the Bohr Model is a modification of the earlier Rutherford Model, some people call Bohr's Model the Rutherford-Bohr Model. The modern model of the atom is based on quantum mechanics. The Bohr Model contains some errors, but it is important because it describes most of the accepted features of atomic theory without all of the high-level math of the modern version. Unlike earlier models, the Bohr Model explains the Rydberg formula for the spectral emission lines of atomic hydrogen.

The Bohr Model is a planetary model in which the negatively-charged electrons orbit a small, positively-charged nucleus similar to the planets orbiting the Sun (except that the orbits are not planar). The gravitational force of the solar system is mathematically akin to the Coulomb (electrical) force between the positively-charged nucleus and the negatively-charged electrons.

Main Points of the Bohr Model

- Electrons orbit the nucleus in orbits that have a set size and energy.

- The energy of the orbit is related to its size. The lowest energy is found in the smallest orbit.

- Radiation is absorbed or emitted when an electron moves from one orbit to another.

Bohr Model of Hydrogen

The simplest example of the Bohr Model is for the hydrogen atom $(Z = 1)$ or for a hydrogen-like ion $(Z > 1)$, in which a negatively-charged electron orbits a small positively-charged nucleus. Electromagnetic energy will be absorbed or emitted if an electron moves from one orbit to another. Only certain electron orbits are permitted. The radius of the possible orbits increases as n^2, where n is the principal quantum number. The $3 \rightarrow 2$ transition produces the first line of the Balmer series. For hydrogen $(Z = 1)$ this produces a photon having wavelength 656 nm (red light).

Bohr Model for Heavier Atoms

Heavier atoms contain more protons in the nucleus than the hydrogen atom. More electrons were required to cancel out the positive charge of all of these protons. Bohr believed each electron orbit could only hold a set number of electrons. Once the level was full, additional electrons would be bumped up to the next level. Thus, the Bohr model for heavier atoms described electron shells. The model explained some of the atomic properties of heavier atoms, which had never been reproduced before. For example, the shell model explained why atoms got smaller moving across a period (row) of the periodic table, even though they had more protons and electrons. It also explained why the noble gases were inert and why atoms on the left side of the periodic table attract electrons, while those on the right side lose them. However, the model assumed electrons in the shells didn't interact with each other and couldn't explain why electrons seemed to stack in an irregular manner.

Problems with the Bohr Model

- It violates the Heisenberg Uncertainty Principle because it considers electrons to have both a known radius and orbit.

- The Bohr Model provides an incorrect value for the ground state orbital angular momentum.

- It makes poor predictions regarding the spectra of larger atoms.

- It does not predict the relative intensities of spectral lines.

- The Bohr Model does not explain fine structure and hyperfine structure in spectral lines.

- It does not explain the Zeeman Effect.

Refinements and Improvements to the Bohr Model

The most prominent refinement to the Bohr model was the Sommerfeld model, which is sometimes called the Bohr-Sommerfeld model. In this model, electrons travel in elliptical orbits around the nucleus rather than in circular orbits. The Sommerfeld model was better at explaining atomic spectral effects, such the Stark effect in spectral line splitting. However, the model couldn't accommodate the magnetic quantum number.

Ultimately, the Bohr model and models based upon it were replaced Wolfgang Pauli's model based on quantum mechanics in 1925. That model was improved to produce the modern model, introduced by Erwin Schrodinger in 1926. Today, the behavior of the hydrogen atom is explained using wave mechanics to describe atomic orbitals.

ATOMIC SPECTRA

When atoms are excited they emit light of certain wavelengths which correspond to different colors. The emitted light can be observed as a series of colored lines with dark spaces in between; this series of colored lines is called a line or atomic spectra. Each element produces a unique set of spectral lines. Since no two elements emit the same spectral lines, elements can be identified by their line spectrum.

Electromagnetic Radiation and the Wave Particle Duality

Energy can travel through a vacuum or matter as electromagnetic radiation. Electromagnetic radiation is a transverse wave with magnetic and electric components that oscillate perpendicular to each other. The electromagnetic spectrum is the range of all possible wavelengths and frequencies of electromagnetic radiation including visible light.

According to the wave particle duality concept, although electromagnetic radiation is often considered to be a wave, it also behaves like a particle. In 1900, while studying black body radiation, Max Planck discovered that energy was limited to certain values and was not continuous as assumed in classical physics. This means that when energy increases, it does so by tiny jumps called quanta (quantum in the singular). In other words, a quantum of energy is to the total energy of a system as an atom is to the total mass of a system. In 1905, Albert Einstein proposed that energy was bundled into packets, which became known as photons. The discovery of photons explained why energy increased in small jumps. If energy was bundled into tiny packets, each additional packet would contribute a tiny amount of energy causing the total amount of energy to jump by a tiny amount, rather than increase smoothly as assumed in classical physics.

- λ is the wavelength of light.

- ν is the frequency of light.

- n is the quantum number of a energy state.

- E is the energy of that state.

Table: Important Constants.

Constant	Meaning	Value
c	speed of light	$2.99792458 \times 10^8\ ms^{-1}$
h	Planck's constant	$6.62607 \times 10^{-34}\ Js$
eV	electron volt	$1.60218 \times 10^{-19}\ J$
R_H	Rydberg constant for H	$2.179 \times 10^{-18}\ J$

Units to Know

Wavelength, or the distance from one peak to the other of a wave, is most often measured in meters, but can be measured using other SI units of length where practical. The number of waves that pass per second is the frequency of the wave. The SI unit for frequency is the Hertz (abbreviated Hz). 1 Hz is equal to $1s^{-1}$. The speed of light is constant. In a vacuum the speed of light is $2.99792458 \times 10^8\ ms^{-1}$. The relationship between wavelength (λ), frequency (ν), and the speed of light (c) is:

$$v = \frac{c}{\lambda}$$

The energy of electromagnetic radiation of a particular frequency is measured in Joules and is given by the equation:

$$E = h\nu$$

with

- h as Planck's constant ($6.62606876\ x\ 10^{-34}\ Js$)

The electron volt is another unit of energy that is commonly used. The electron volt (eV) is defined as the kinetic energy gained by an electron when it is accelerated by a potential electrical difference of 1 volt. It is equal to $1.60218\ x\ 10^{-19}\ J$.

Spectroscope

A spectrum is a range of frequencies or wavelengths. By the process of refraction, a prism can split white light into it's component wavelengths. However this method is rather crude, so a spectroscope is used to analyze the light passing through the prism more accurately. The diagram to the right shows a simple prism spectroscope. The smaller the difference between distinguishable wavelengths, the higher the resolution of the spectroscope. The observer sees the radiation passing through the slit as a spectral line. To obtain accurate measurements of the radiation, and electronic device often takes the place of the observer, the device is then called a spectrophotometer. In more modern Spectrophotometers, a diffraction grating is used instead of a prism to disperse the light.

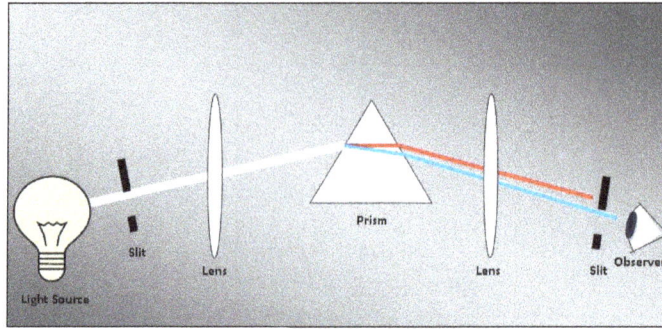

How Atoms React when Excited by Light

Electrons can only exist in certain areas around the nucleus called shells. Each shell corresponds to a specific energy level which is designated by a quantum number n. Since electrons cannot exist between energy levels, the quantum number n is always an integer value $(n = 1,2,3,4...)$. The electron with the lowest energy level $(n = 1)$ is the closest to the nucleus. An electron occupying its lowest energy level is said to be in the ground state. The energy of an electron in a certain energy level can be found by the equation:

$$E_n = \frac{-R_H}{n^2}$$

Where R_H is a constant equal to $2.179 \ x \ 10^{-18} \ J$ and n is equal to the energy level of the electron.

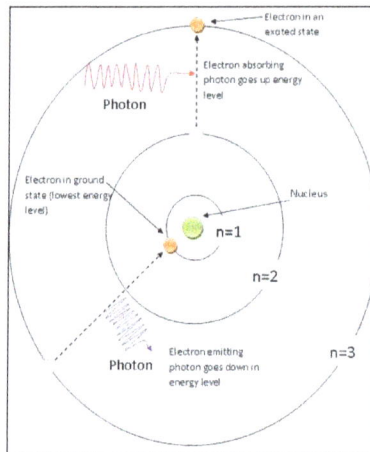

When light is shone on an atom, its electrons absorb photons which cause them to gain energy and jump to higher energy levels. The higher the energy of the photon absorbed, the higher the energy level the electron jumps to. Similarly, an electron can go down energy levels by emitting a photon. The simplified version of this principal is illustrated in the figure to the left based on the Bohr model of the Hydrogen atom. The energy of the photon emitted or gained by an electron can be calculated from this formula:

$$E_{photon} = R_H \left(\frac{1}{n_i^2} - \frac{1}{n_f^2} \right)$$

Where n_i is the initial energy level of the electron and n_f is the final energy level of the electron. The frequency of the photon emitted when an electron descends energy levels can be found using the formula:

$$nu_{photon} = \frac{E_i - E_f}{h}$$

with

- E_i is the initial energy of the electron.

- E_f is the final energy of the electron.

Since an electron can only exist at certain energy levels, they can only emit photons of certain frequencies. These specific frequencies of light are then observed as spectral lines. Similarly, a photon has to be of the exact wavelength the electron needs to jump energy levels in order to be absorbed, explaining the dark bands of an absorption spectra.

Emission Lines

When an electron falls from one energy level in an atom to a lower energy level, it emits a photon of a particular wavelength and energy. When many electrons emit the same wavelength of photons it will result in a spike in the spectrum at this particular wavelength, resulting in the banding pattern seen in atomic emission spectra. The graphic to the right is a simplified picture of a spectrograph, in this case being used to photograph the spectral lines of Hydrogen.

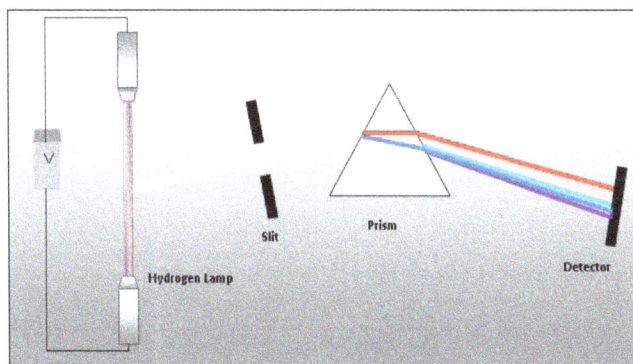

In this spectrograph, the Hydrogen atoms inside the lamp are being excited by an electric current. The light from the lamp then passes through a prism, which diffracts it into its different frequencies. Since the frequencies of light correspond to certain energy levels (n) it is therefore possible to predict the frequencies of the spectral lines of Hydrogen using an equation discovered by Johann Balmer.

$$v = 3.2881x10^{15}\,s^{-1}\left(\frac{1}{2^2} - \frac{1}{n^2}\right)$$

Where n must be a number greater than 2. This is because Balmer's formula only applies to visible light and some longer wavelengths of ultraviolet.

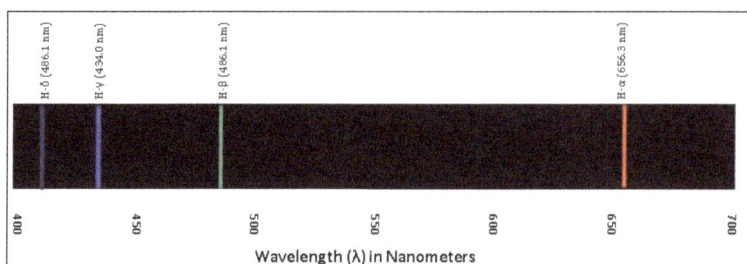

Balmer Series for Hydrogen Atom.

The frequencies in this region of Hydrogen's atomic spectra are called the Balmer series. The Balmer series for Hydrogen is pictured to the left. There are several other series in the Hydrogen atom which correspond to different parts of the electromagnetic spectrum. The Lyman series, for example, extends into the ultraviolet, and therefore can be used to calculate the energy of to $n = 1$.

Absorption Lines

When an electron jumps from a low energy level to a higher level, the electron will absorb a photon of a particular wavelength. This will show up as a drop in the number of photons of this wavelength and as a black band in this part of the spectrum. The figure to the right illustrates a mechanism to detect an absorption spectrum. A white light is shone through a sample. The atoms in the sample absorb some of the light, exciting their electrons. Since the electrons only absorb light of certain frequencies, the absorption spectrum will show up as a series of black bands on an otherwise continuous spectrum.

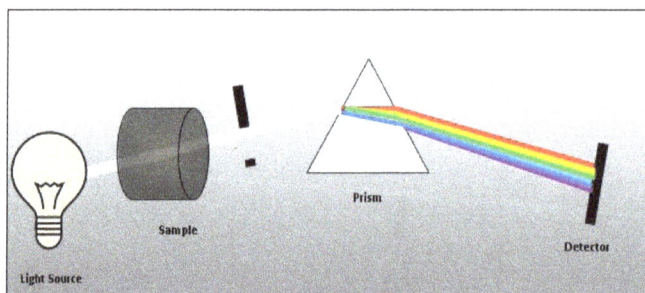

Applications of Atomic Spectral Analysis

Atomic spectroscopy has many useful applications. Since the emission spectrum is different for every element, it acts as an atomic fingerprint by which elements can be identified. Some elements were discovered by the analysis of their atomic spectrum. Helium, for example, was discovered while scientists were analyzing the absorption spectrum of the sun. Emission spectra is especially useful to astronomers who use emission and absorption spectra to determine the make up of far away stars and other celestial bodies.

BOHR MODEL OF HYDROGEN

Niels Bohr introduced the atomic Hydrogen model in 1913. He described it as a positively charged nucleus, comprised of protons and neutrons, surrounded by a negatively charged electron cloud.

In the model, electrons orbit the nucleus in atomic shells. The atom is held together by electrostatic forces between the positive nucleus and negative surroundings.

Hydrogen Energy Levels

The Bohr model is used to describe the structure of hydrogen energy levels. The image below represents shell structure, where each shell is associated with principal quantum number n. The energy levels presented correspond with each shell. The amount of energy in each level is reported in eV, and the maxiumum energy is the ionization energy of 13.598eV.

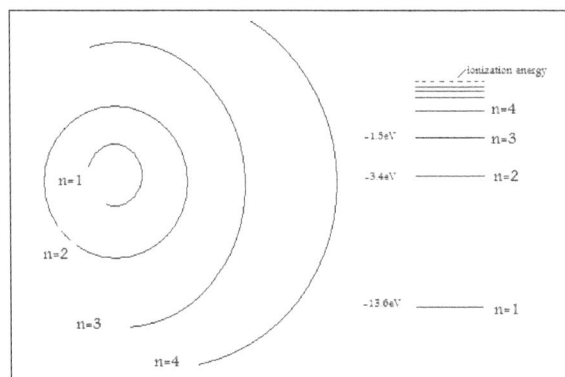

Some of the orbital shells of a Hydrogen atom. The energy levels of the orbitals are shown to the right.

Hydrogen Spectrum

The movement of electrons between these energy levels produces a spectrum. The Balmer equation is used to describe the four different wavelengths of Hydrogen which are present in the visible light spectrum. These wavelengths are at 656, 486, 434, and 410nm. These correspond to the emission of photons as an electron in an excited state transitions down to energy level n = 2. The Rydberg formula, below, generalizes the Balmer series for all energy level transitions. To get the Balmer lines, the Rydberg formula is used with an n_f of 2.

Rydberg Formula

The Rydberg formula explains the different energies of transition that occur between energy levels. When an electron moves from a higher energy level to a lower one, a photon is emitted. The Hydrogen atom can emit different wavelengths of light depending on the initial and final energy levels of the transition. It emits a photon with energy equal to the difference of square of the final $\left(n_f \right)$ and initial $\left(n_i \right)$ energy levels.

$$\text{Energy} = R\left(\frac{1}{n_f^2} - \frac{1}{n_i^2} \right)$$

The energy of a photon is equal to Planck's constant, $h = 6.626*10^{-34} m^2 kg / s$, times the speed of light in a vacuum, divided by the wavelength of emission.

$$E = \frac{hc}{\lambda}$$

Combining these two equations produces the Rydberg Formula.

$$\frac{1}{\lambda} = R\left(\frac{1}{n_f^2} - \frac{1}{n_i^2}\right)$$

he Rydberg Constant (R) $=10,973,731.6m^{-1}$ or $1.097 \times 107 m^{-1}$.

HYDROGEN SPECTRUM

Bohr was able to derive the formula for the hydrogen spectrum using basic physics, the planetary model of the atom, and some very important new proposals. His first proposal is that only certain orbits are allowed: we say that the orbits of electrons in atoms are quantized. Each orbit has a different energy, and electrons can move to a higher orbit by absorbing energy and drop to a lower orbit by emitting energy. If the orbits are quantized, the amount of energy absorbed or emitted is also quantized, producing discrete spectra. Photon absorption and emission are among the primary methods of transferring energy into and out of atoms. The energies of the photons are quantized, and their energy is explained as being equal to the change in energy of the electron when it moves from one orbit to another. In equation form, this is:

$$\Delta E = hf = E_i - E_f.$$

Here, ΔE is the change in energy between the initial and final orbits, and hf is the energy of the absorbed or emitted photon. It is quite logical (that is, expected from our everyday experience) that energy is involved in changing orbits. A blast of energy is required for the space shuttle, for example, to climb to a higher orbit. What is not expected is that atomic orbits should be quantized. This is not observed for satellites or planets, which can have any orbit given the proper energy.

The planetary model of the atom, as modified by Bohr, has the orbits of the electrons quantized. Only certain orbits are allowed, explaining why atomic spectra are discrete (quantized). The energy carried away from an atom by a photon comes from the electron dropping from one allowed orbit to another and is thus quantized. This is likewise true for atomic absorption of photons.

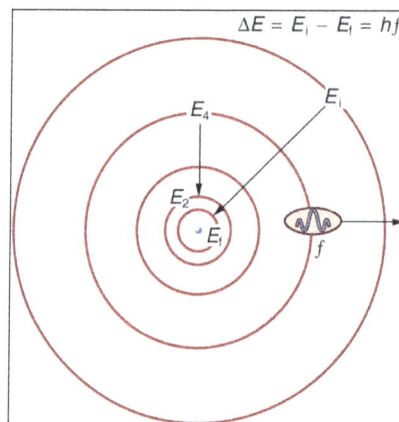

Energy is plotted vertically with the lowest or ground state at the bottom and with excited states above. Given the energies of the lines in an atomic spectrum, it is possible (although sometimes very difficult) to determine the energy levels of an atom. Energy-level diagrams are used for many systems, including molecules and nuclei. A theory of the atom or any other system must predict its energies based on the physics of the system.

An energy-level diagram plots energy vertically and is useful in visualizing the energy states of a system and the transitions between them. This diagram is for the hydrogen-atom electrons, showing a transition between two orbits having energies E_4 and E_2.

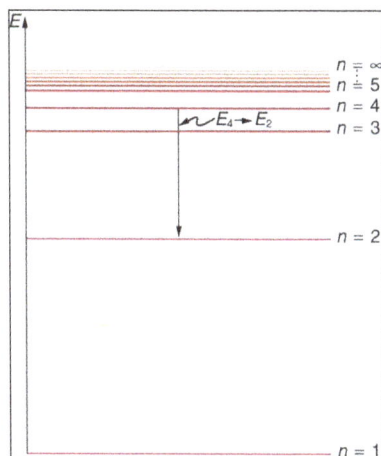

Bohr was clever enough to find a way to calculate the electron orbital energies in hydrogen. This was an important first step that has been improved upon, but it is well worth repeating here, because it does correctly describe many characteristics of hydrogen. Assuming circular orbits, Bohr proposed that the angular momentum L of an electron in its orbit is quantized, that is, it has only specific, discrete values. The value for L is given by the formula:

$$L = m_e \mathrm{vr}_n = n\frac{h}{2\pi}\left(n = 1, 2, 3, \ldots\right),$$

where L is the angular momentum, m_e is the electron's mass, r_n is the radius of the n th orbit, and h is Planck's constant. Note that angular momentum is $L = I\omega$. For a small object at a radius $r, I = mr^2$ and $\omega = v/r$, so that $L = \left(mr^2\right)\left(v/r\right) = mvr$. Quantization says that this value of mvr can only be equal to $h/2$, $2h/2$, $3h/2$, etc. At the time, Bohr himself did not know why angular momentum should be quantized, but using this assumption he was able to calculate the energies in the hydrogen spectrum, something no one else had done at the time.

From Bohr's assumptions, we will now derive a number of important properties of the hydrogen atom from the classical physics we have covered in the text. We start by noting the centripetal force causing the electron to follow a circular path is supplied by the Coulomb force. To be more general, we note that this analysis is valid for any single-electron atom. So, if a nucleus has Z protons (Z = 1 for hydrogen, 2 for helium, etc.) and only one electron, that atom is called a hydrogen-like atom. The spectra of hydrogen-like ions are similar to hydrogen, but shifted to higher energy by the greater attractive force between the electron and nucleus. The magnitude of the centripetal force is $m_e v^2 / r_n$, while the Coulomb force is $k\left(Zq_e\right)\left(q_e\right)/r_n^2$. The tacit assumption here is that the nucleus is more massive than the stationary electron, and the electron orbits about it. This is consistent with the planetary model of the atom. Equating these,

$$k\frac{Zq_e^2}{r_n^2} = \frac{m_e v^2}{r_n}\ (Coulomb = centripetal).$$

Angular momentum quantization is stated in an earlier equation. We solve that equation for v, substitute it into the above, and rearrange the expression to obtain the radius of the orbit. This yields:

$$r_n = \frac{n^2}{Z}a_B,\ \text{for allowed orbits}\ (n = 1,2,3,\ldots),$$

where a_B is defined to be the Bohr radius, since for the lowest orbit $(n = 1)$ and for hydrogen $(Z = 1), r_1 = a_B$. It is left for this chapter's Problems and Exercises to show that the Bohr radius is:

$$a_B = \frac{h^2}{4\pi^2 m_e kq_e^2} = 0.52910^{-10}m$$

These last two equations can be used to calculate the radii of the allowed (quantized) electron orbits in any hydrogen-like atom. It is impressive that the formula gives the correct size of hydrogen, which is measured experimentally to be very close to the Bohr radius. The earlier equation also tells us that the orbital radius is proportional to n^2, as illustrated below.

The allowed electron orbits in hydrogen have the radii shown. These radii were first calculated by Bohr and are given by the equation $r_n = \frac{n^2}{Z}a_B$. The lowest orbit has the experimentally verified diameter of a hydrogen atom.

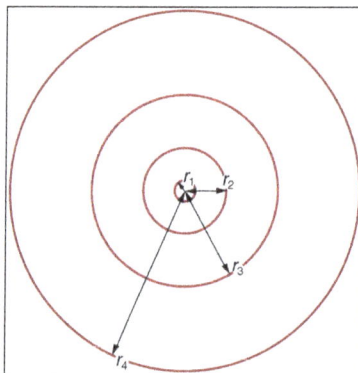

To get the electron orbital energies, we start by noting that the electron energy is the sum of its kinetic and potential energy:

$$E_n = KE + PE.$$

Kinetic energy is the familiar $KE = (1/2)m_e v^2$, assuming the electron is not moving at relativistic speeds. Potential energy for the electron is electrical, or $PE = q_e V$, where V is the potential due to the nucleus, which looks like a point charge. The nucleus has a positive charge Zq_e; thus, $V = kZq_e / r_n$, recalling an earlier equation for the potential due to a point charge. Since the electron's charge is negative, we see that $PE = -kZq_e / r_n$. Entering the expressions for KE *and* PE, we find:

$$E_n = \frac{1}{2} m_e v^2 - k \frac{Zq_e^2}{r_n}.$$

Now we substitute r_n and v from earlier equations into the above expression for energy. Algebraic manipulation yields:

$$E_n = -\frac{Z^2}{n^2} E_0 \left(n = 1, 2, 3, \ldots \right)$$

for the orbital energies of hydrogen-like atoms. Here, E_0 is the ground-state energy $\left(n = 1 \right)$ for hydrogen $\left(Z = 1 \right)$ and is given by:

$$E_0 = \frac{2\pi^2 q_e^4 m_e k^2}{h^2} = 13.6 \text{ eV}.$$

Thus, for hydrogen,

$$E_n = -\frac{13.6 \text{ eV}}{n^2} (n = 1, 2, 3, \ldots).$$

for hydrogen that also illustrates how the various spectral series for hydrogen are related to transitions between energy levels.

Energy-level diagram for hydrogen showing the Lyman, Balmer, and Paschen series of transitions. The orbital energies are calculated using the above equation, first derived by Bohr.

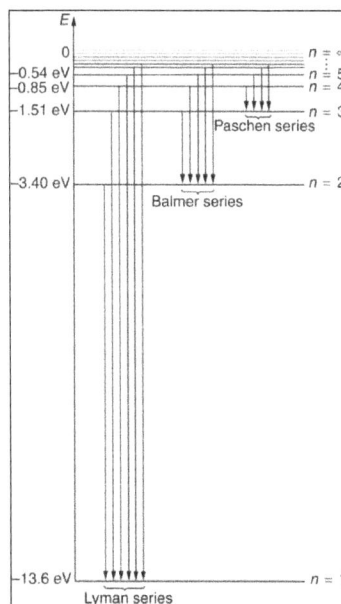

Electron total energies are negative, since the electron is bound to the nucleus, analogous to being in a hole without enough kinetic energy to escape. As n approaches infinity, the total energy becomes zero. This corresponds to a free electron with no kinetic energy, since r_n gets very large for large n, and the electric potential energy thus becomes zero. Thus, 13.6 eV is needed to ionize hydrogen (to go from −13.6 eV to 0, or unbound), an experimentally verified number. Given more energy, the electron becomes unbound with some kinetic energy. For example, giving 15.0 eV to an electron in the ground state of hydrogen strips it from the atom and leaves it with 1.4 eV of kinetic energy.

Finally, let us consider the energy of a photon emitted in a downward transition, given by the equation to be:

$$\Delta E = hf = E_i - E_f.$$

Substituting $E_n = \left(-13.6 \text{ eV} / n^2 \right)$, we see that:

$$hf = \left(13.6 \text{ eV} \right) \left(\frac{1}{n_f^2} - \frac{1}{n_i^2} \right).$$

Dividing both sides of this equation by hc gives an expression for $1 / \lambda$:

$$\frac{hf}{hc} = \frac{f}{c} = \frac{1}{\lambda} = \frac{\left(13.6 \text{ eV} \right)}{hc} \left(\frac{1}{n_f^2} - \frac{1}{n_i^2} \right).$$

It can be shown that:

$$\left(\frac{13.6 \text{ eV}}{hc} \right) = \frac{\left(13.6 \text{eV} \right) \left(1.602 10^{-19} \text{J/eV} \right)}{\left(6.626 10^{-34} \text{Js} \right) \left(2.998 10^8 \text{m/s} \right)} = 1.097 10^7 \text{m}^{-1} = R$$

is the Rydberg constant. Thus, we have used Bohr's assumptions to derive the formula first proposed by Balmer years earlier as a recipe to fit experimental data.

$$\frac{1}{\lambda} = R \left(\frac{1}{n_f^2} - \frac{1}{n_i^2} \right)$$

We see that Bohr's theory of the hydrogen atom answers the question as to why this previously known formula describes the hydrogen spectrum. It is because the energy levels are proportional to $1 / n^2$, where n is a non-negative integer. A downward transition releases energy, and so n_i must be greater than n_f. The various series are those where the transitions end on a certain level. For the Lyman series, $n_f = 1$ — that is, all the transitions end in the ground state. For the Balmer series, $n_f = 2$, or all the transitions end in the first excited state; and so on. What was once a recipe is now based in physics, and something new is emerging—angular momentum is quantized.

Triumphs and Limits of the Bohr Theory

Bohr did what no one had been able to do before. Not only did he explain the spectrum of hydrogen, he correctly calculated the size of the atom from basic physics. Some of his ideas are broadly applicable. Electron orbital energies are quantized in all atoms and molecules. Angular momentum is quantized. The electrons do not spiral into the nucleus, as expected classically (accelerated charges radiate, so that the electron orbits classically would decay quickly, and the electrons would sit on the nucleus—matter would collapse). These are major triumphs.

But there are limits to Bohr's theory. It cannot be applied to multielectron atoms, even one as simple as a two-electron helium atom. Bohr's model is what we call semiclassical. The orbits are quantized (nonclassical) but are assumed to be simple circular paths (classical). As quantum mechanics was developed, it became clear that there are no well-defined orbits; rather, there are clouds of probability. Bohr's theory also did not explain that some spectral lines are doublets (split into two) when examined closely. We shall examine many of these aspects of quantum mechanics in more detail, but it should be kept in mind that Bohr did not fail. Rather, he made very important steps along the path to greater knowledge and laid the foundation for all of atomic physics that has since evolved.

FRANK-HERTZ EXPERIMENT

The Franck–Hertz experiment was the first electrical measurement to clearly show the quantum nature of atoms, and thus "transformed our understanding of the world". It was presented on April 24, 1914, to the German Physical Society in a paper by James Franck and Gustav Hertz. Franck and Hertz had designed a vacuum tube for studying energetic electrons that flew through a thin vapor of mercury atoms. They discovered that, when an electron collided with a mercury atom, it could lose only a specific quantity (4.9 electron volts) of its kinetic energy before flying away. This energy loss corresponds to decelerating the electron from a speed of about 1.3 million meters per second to zero. A faster electron does not decelerate completely after a collision, but loses precisely the same amount of its kinetic energy. Slower electrons merely bounce off mercury atoms without losing any significant speed or kinetic energy.

These experimental results proved to be consistent with the Bohr model for atoms that had been proposed the previous year by Niels Bohr. The Bohr model was a precursor of quantum mechanics and of the electron shell model of atoms. Its key feature was that an electron inside an atom occupies one of the atom's "quantum energy levels". Before the collision, an electron inside the mercury atom occupies its lowest available energy level. After the collision, the electron inside occupies a higher energy level with 4.9 electron volts (eV) more energy. This means that the electron is more loosely bound to the mercury atom. There were no intermediate levels or possibilities in Bohr's quantum model. This feature was "revolutionary" because it was inconsistent with the expectation that an electron could be bound to an atom's nucleus by any amount of energy.

In a second paper presented in May 1914, Franck and Hertz reported on the light emission by the mercury atoms that had absorbed energy from collisions. They showed that the wavelength of this ultraviolet light corresponded exactly to the 4.9 eV of energy that the flying electron had lost. The relationship of energy and wavelength had also been predicted by Bohr. After a presentation of these results by Franck a few years later, Albert Einstein is said to have remarked, "It's so lovely it makes you cry."

On December 10, 1926, Franck and Hertz were awarded the 1925 Nobel Prize in Physics "for their discovery of the laws governing the impact of an electron upon an atom".

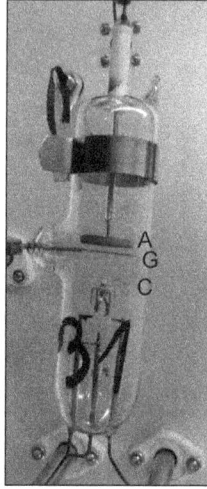

Photograph of a vacuum tube used for the Franck–Hertz experiment in instructional laboratories. There is a droplet of mercury inside the tube, although it is not visible in the photograph. C - cathode assembly; the cathode itself is hot, and glows orange. It emits electrons which pass through the metal mesh grid (G) and are collected as an electric current by the anode (A).

Experiment

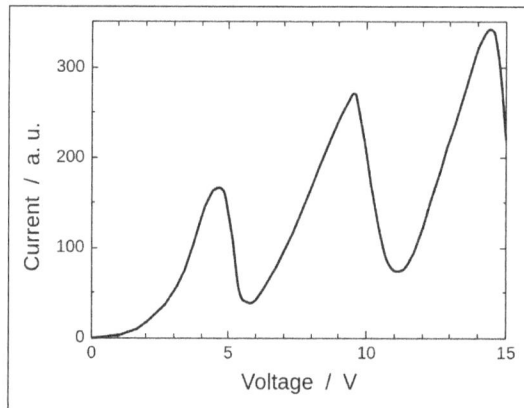

Anode current (arbitrary units) versus grid voltage (relative to the cathode).
This graph is based on the original 1914 paper by Franck and Hertz.

Franck and Hertz's original experiment used a heated vacuum tube containing a drop of mercury; they reported a tube temperature of 115 °C, at which the vapor pressure of mercury is about 100 pascals (and far below atmospheric pressure). A contemporary Franck–Hertz tube is shown in the photograph. It is fitted with three electrodes: an electron-emitting, hot cathode; a metal mesh grid; and an anode. The grid's voltage is positive relative to the cathode, so that electrons emitted from the hot cathode are drawn to it. The electric current measured in the experiment is due to electrons that pass through the grid and reach the anode. The anode's electric potential is slightly negative relative to the grid, so that electrons that reach the anode have at least a corresponding amount of kinetic energy after passing the grid.

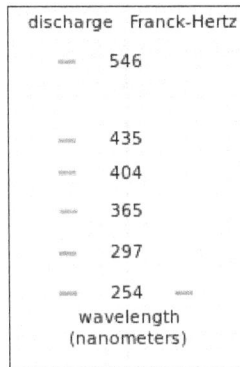

Wavelengths of light emitted by a mercury vapor discharge and by a Franck–Hertz tube in operation at 10 V. The Franck–Hertz tube primarily emits light with a wavelength near 254 nanometers; the discharge emits light at many wavelengths. Based on the original 1914 figure.

The graphs published by Franck and Hertz show the dependence of the electric current flowing out of the anode upon the electric potential between the grid and the cathode.

- At low potential differences—up to 4.9 volts—the current through the tube increased steadily with increasing potential difference. This behavior is typical of true vacuum tubes that don't contain mercury vapor; larger voltages lead to larger "space-charge limited current".

- At 4.9 volts the current drops sharply, almost back to zero.

- The current then increases steadily once again as the voltage is increased further, until 9.8 volts is reached (exactly 4.9+4.9 volts).

- At 9.8 volts a similar sharp drop is observed.

- While it isn't evident in the original measurements of the figure, this series of dips in current at approximately 4.9 volt increments continues to potentials of at least 70 volts.

Franck and Hertz noted in their first paper that the 4.9 eV characteristic energy of their experiment corresponded well to one of the wavelengths of light emitted by mercury atoms in gas discharges. They were using a quantum relationship between the energy of excitation and the corresponding wavelength of light, which they broadly attributed to Johannes Stark and to Arnold Sommerfeld; it predicts that 4.9 eV corresponds to light with a 254 nm wavelength. The same relationship was also incorporated in Einstein's 1905 photon theory of the photoelectric effect. In a second paper, Franck and Hertz reported the optical emission from their tubes, which emitted light with a single prominent wavelength 254 nm. The figure at the right shows the spectrum of a Franck–Hertz tube; nearly all of the light emitted has a single wavelength. For reference, the figure also shows the spectrum for a mercury gas discharge light, which emits light at several wavelengths besides 254 nm. The figure is based on the original spectra published by Franck and Hertz in 1914. The fact that the Franck–Hertz tube emitted just the single wavelength, corresponding nearly exactly to the voltage period they had measured, was very important.

Modeling of Electron Collisions with Atoms

Franck and Hertz explained their experiment in terms of elastic and inelastic collisions between the electrons and the mercury atoms. Slowly moving electrons collide elastically with the mercury atoms. This means that the direction in which the electron is moving is altered by the collision, but its speed is unchanged. An elastic collision is illustrated in the figure, where the length of the arrow indicates the electron's speed. The mercury atom is unaffected by the collision, mostly because it is about four hundred thousand times more massive than an electron.

When the speed of the electron exceeds about 1.3 million meters per second, collisions with a mercury atom become inelastic. This speed corresponds to a kinetic energy of 4.9 eV, which is deposited into the mercury atom. The electron's speed is reduced, and the mercury atom becomes "excited". A short time later, the 4.9 eV of energy that was deposited into the mercury atom is released as ultraviolet light that has a wavelength of precisely 254 nm. Following light emission, the mercury atom returns to its original, unexcited state.

If electrons emitted from the cathode flew freely until they arrived at the grid, they would acquire a kinetic energy that's proportional to the voltage applied to the grid. 1 eV of kinetic energy corresponds to a potential difference of 1 volt between the grid and the cathode. Elastic collisions with the mercury atoms increase the time it takes for an electron to arrive at the grid, but the average kinetic energy of electrons arriving there isn't much affected.

When the grid voltage reaches 4.9 V, electron collisions near the grid become inelastic, and the electrons are greatly slowed. The kinetic energy of a typical electron arriving at the grid is reduced so much that it cannot travel further to reach the anode, whose voltage is set to slightly repel electrons. The current of electrons reaching the anode falls, as seen in the graph. Further increases in the grid voltage restore enough energy to the electrons that suffered inelastic collisions that they can again reach the anode. The current rises again as the grid potential rises beyond 4.9 V. At 9.8 V, the situation changes again. Electrons that have traveled roughly halfway from the cathode to the grid have already acquired enough energy to suffer a first inelastic collision. As they continue slowly towards the grid from the midway point, their kinetic energy builds up again, but as they reach the grid they can suffer a second inelastic collision. Once again, the current to the anode drops. At intervals of 4.9 volts this process will repeat; each time the electrons will undergo one additional inelastic collision.

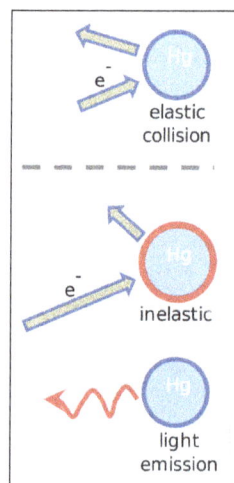

Elastic and inelastic collisions of electrons with mercury atoms. Electrons traveling slowly change direction after elastic collisions, but do not change their speed. Faster electrons lose most of their speed in inelastic collisions. The lost kinetic energy is deposited into the mercury atom. The atom subsequently emits light, and returns to its original state.

Early Quantum Theory

While Franck and Hertz were unaware of it when they published their experiments in 1914, in 1913 Niels Bohr had published a model for atoms that was very successful in accounting for the optical properties of atomic hydrogen. These were usually observed in gas discharges, which emitted light at a series of wavelengths. Ordinary light sources like incandescent light bulbs emit light at all wavelengths. Bohr had calculated the wavelengths emitted by hydrogen very accurately.

The fundamental assumption of the Bohr model concerns the possible binding energies of an electron to the nucleus of an atom. The atom can be ionized if a collision with another particle supplies at least this binding energy. This frees the electron from the atom, and leaves a positively charged ion behind. There is an analogy with satellites orbiting the earth. Every satellite has its own orbit, and practically any orbital distance, and any satellite binding energy, is possible. Since an electron is attracted to the positive charge of the atomic nucleus by a similar force, so-called "classical" calculations suggest that any binding energy should also be possible for electrons. However, Bohr assumed that only a specific series of binding energies occur, which correspond to the "quantum energy levels" for the electron. An electron is normally found in the lowest energy level, with the largest binding energy. Additional levels lie higher, with smaller binding energies. Intermediate binding energies lying between these levels are not permitted. This was a revolutionary assumption.

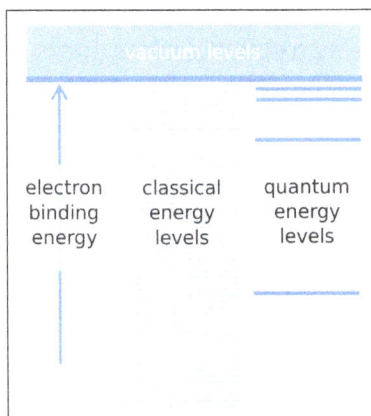

The Bohr model of the atom assumed that an electron could be bound to an atomic nucleus only with one of a series of specific energies corresponding to quantum energy levels. Earlier, classical models for the binding of particles allowed any binding energy.

Franck and Hertz had proposed that the 4.9 V characteristic of their experiments was due to ionization of mercury atoms by collisions with the flying electrons emitted at the cathode. In 1915 Bohr published a paper noting that the measurements of Franck and Hertz were more consistent with the assumption of quantum levels in his own model for atoms. In the Bohr model, the collision excited an internal electron within the atom from its lowest level to the first quantum level above it. The Bohr model also predicted that light would be emitted as the internal electron returned from its excited quantum level to the lowest one; its wavelength corresponded to the energy difference of the atom's internal levels, which has been called the Bohr relation. Franck and Hertz's observation of emission from their tube at 254 nm was also consistent with Bohr's perspective. Writing following the end of World War I in 1918, Franck and Hertz had largely adopted the Bohr perspective for interpreting their experiment, which has become one of the experimental pillars of quantum mechanics. As Abraham Pais described it, "Now the beauty of Franck and Hertz's work lies not only in the measurement of the energy loss $E_2 - E_1$ of the impinging electron, but they also observed that, when the energy of that electron exceeds 4.9 eV, mercury begins to emit ultraviolet light of a definite frequency v as defined in the above formula. Thereby they gave (unwittingly at first) the first direct experimental proof of the Bohr relation!" Franck himself emphasized the importance of the ultraviolet emission experiment in an epilogue to the 1960 Physical Science Study Committee (PSSC) film about the Franck–Hertz experiment.

Experiment with Neon

In instructional laboratories, the Franck–Hertz experiment is often done using neon gas, which shows the onset of inelastic collisions with a visible orange glow in the vacuum tube, and which also is non-toxic, should the tube be broken. With mercury tubes, the model for elastic and inelastic collisions predicts that there should be narrow bands between the anode and the grid where the mercury emits light, but the light is ultraviolet and invisible. With neon, the Franck–Hertz voltage interval is 18.7 volts, and an orange glow appears near the grid when 18.7 volts is applied. This glow will move closer to the cathode with increasing accelerating potential, and indicates the locations where electrons have acquired the 18.7 eV required to excite a neon atom. At 37.4 volts two distinct glows will be visible: one midway between the cathode and grid, and one right at the accelerating grid. Higher potentials, spaced at 18.7 volt intervals, will result in additional glowing regions in the tube.

An additional advantage of neon for instructional laboratories is that the tube can be used at room temperature. However, the wavelength of the visible emission is much longer than predicted by the Bohr relation and the 18.7 V interval. A partial explanation for the orange light involves two atomic levels lying 16.6 eV and 18.7 eV above the lowest level. Electrons excited to the 18.7 eV level fall to the 16.6 eV level, with concomitant orange light emission.

Franck-Hertz experiment with neon gas: 3 glowing regions.

SOMMERFELD MODEL OF THE ATOM

In order to explain the observed fine structure of spectral lines, Sommerfeld introduced two main modifications in Bohr's theory:

- According to Sommerfeld, the path of an electron around the nucleus, in general, is an ellipse with the nucleus at one of its foci.

- The velocity of the electron moving in an elliptical orbit varies at different parts of the orbit. This causes the relativistic variation in the mass of the moving electron.

- Now, when elliptical orbits are permitted, one has to deal with two variable quantities.

- The varying distance of the electron from the nucleus (r).

- The varying angular position of the electron with respect to the nucleus i.e, the azimuthal angle φ in figure.

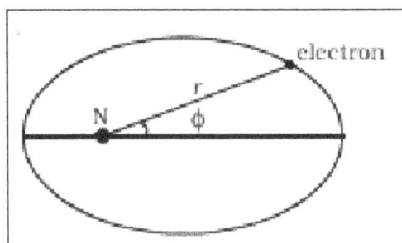

Sommerfeld atom model.

To deal with these two variables, two quantum numbers are introduced:

- The principal quantum number n of Bohr's theory, which determines the energy of the electrons.

- A new quantum number called orbital (or azimuthal) quantum number (l) which has been introduced to characterize the angular momentum in an orbit i.e., it determines the orbital angular momentum of the electron. Its values vary from zero to (n - 1) in steps of unity.

This orbital quantum number (l) is useful in finding the possible elliptical orbits. The possible elliptical orbits are such that:

$$b/a = 1 + 1/n$$

where a and b are semi-major and semi-minor axes respectively of the ellipse.

According to Sommerfeld's model, for any principal quantum number n, there are n possible orbits of varying eccentricities called sub-orbits or sub-shells. Out of n subshells, one is circular and the remaining (i.e., n - 1) are elliptical in shape.

These possible sub-orbits possess slightly different energies because of the relativistic variation of the electron mass.

Consider the first energy level $(n = 1)$. When $n = 1$, $l = 0$ i.e., in this energy level, there is only one orbit or sub-shell for the electron. Also, when a = b, the two axes of the ellipse are equal. As a result of this, the orbit corresponding to $n = 1$ is circular. This subshell is designated as s sub-shell. Since, this sub-shell belongs to $n = 1$, it is designated as 1s in figure.

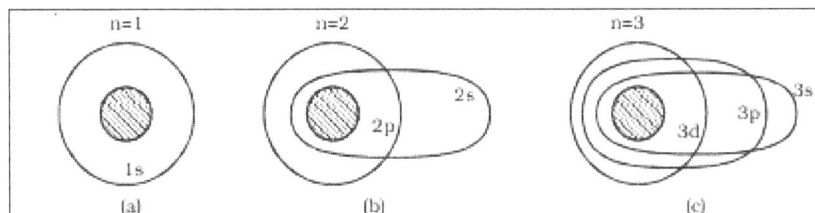

Various sub-shells for the electrons.

Similarly, for the second energy level $n = 2$, there are two permissible sub-shells for the electrons. For $n = 2$, l can take two values, 0 and 1.

When $n = 2$, $l = 0$.

$$b/a = 0+1/2 = 1/2$$

or:

b = a/2

This subshell corresponding to $l = 0$ is elliptical in shape and is designated as 2s.

when $n = 2$, $l = 1$.

$$b/a = 1+1/2 = 2/2 = 1$$

or:

b = a

This sub-shell corresponding to $l = 1$ is circular in shape and is designated as 2p.

For $n = 3$, l has three values 0, 1 and 2, i.e. there are three permissible sub-shells for the electrons.

when $n = 3$, $l = 0$.

$$b/a = (0+1)/3 = 1/3 = 1 or\ b = a/3$$

when $n = 3$, $l = 1$.

$$b/a = (1+1)/3 = 2/3 = 1 or\ b = 2a/3$$

and when $n = 3$, $l = 2$.

$$b/a = (2+1)/3 = 3/3 = 1 or\ b = a$$

The sub-shells corresponding to $l = 0$, 1 and 2 are designated as 3s, 3p and 3d respectively. The circular shell is designated as 3d and the other two are elliptical in shape.

It is common practice to assign letters to l-values as given below:

Orbital quantum number l: 0   1   2   3   4

electron state: s   p   d   f   g

Hence, electrons in the $l = 0$, 1, 2, 3❖ states are said to be in the s, p, d, f❖ states.

Fine Structure of Spectral Line

Based on Sommerfeld atom model, the total energy of an electron in the elliptical orbit can be shown as,

$$En = (-me^4Z^2) / (8\varepsilon_0^2 h^2 n^2)$$

This expression is the same as that obtained by Bohr. Thus the introduction of elliptical orbits gives no new energy levels and hence no new transition. In this way, the attempt of Sommerfeld to explain the fine structure of spectral lines failed. But soon, on the basis of variation of mass of electron with velocity, Sommerfeld could find the solution for the problem of the fine structure of the spectral lines.

According to Sommerfeld, the velocity of the electron is maximum when the electron is nearest to the nucleus and minimum when it is farthest from the nucleus, since the orbit of the electron is elliptical. This implies that the effective mass of the electron will be different at different parts of its orbit. Taking into account the relativistic variation of the mass of the electron, Sommerfeld modified his theory and showed that the path of electron is not a simple ellipse but a precessing ellipse called a rosette, in figure.

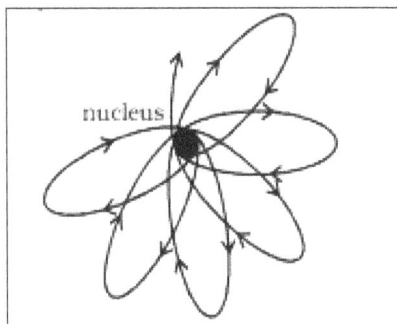

Rosette path of an electron.

Based on this idea, Sommerfeld successfully explained the fine structure of spectral lines of hydrogen atom.

Drawbacks

- Though Sommerfeld's modification gave a theoretical background of the fine structure of spectral lines of hydrogen, it could not predict the correct number of observed fine structure of these lines.

- It could not explain the distribution and arrangement of electrons in atoms.

- Sommerfeld's model was unable to explain the spectra of alkali metals such as sodium, potassium etc.

- It could not explain Zeeman and Stark effect.

- This model does not give any explanation for the intensities of the spectral lines.

References

- Thornton, Stephen; Rex, Andrew (2012). Modern Physics for Scientists and Engineers (4 ed.). Cengage Learning. Pp. 154–156. ISBN 9781133103721

- Bohr-model-of-the-atom: thoughtco.com, Retrieved 21 May, 2019

- Atomic-Spectra, Electrons-in-Atoms, The-Hydrogen-Atom, Quantum-Mechanics, Supplemental-Modules-(Physical-and-Theoretical-Chemistry), Physical-and-Theoretical-Chemistry-Textbook-Maps, Bookshelves: libretexts.org, Retrieved 17 February, 2019

- For converting electron volts to electron speeds, see "The speed of electrons". Practical Physics. Nuffield Foundation. Retrieved 2014-04-18

- Bohrs-theory-of-the-hydrogen-atom, chapter, physicstestbook2: opentextbc.ca, Retrieved 24 March, 2019

- Sommerfeld-atom-model-and-its-Drawbacks: brainkart.com, Retrieved 21 April, 2019

Quantum Mechanical Model of the Atom

Quantum mechanical model is the model of an atom which is based on mathematics and quantum theory. It is primarily used to explain the observations made on complex atoms. This chapter closely examines the key concepts of quantum mechanical model of atom to provide an extensive understanding of the subject.

In 1926, Austrian physicist Erwin Schrödinger used the wave-particle duality of the electron to develop and solve a complex mathematical equation that accurately described the behavior of the electron in a hydrogen atom. The quantum mechanical model of the atom comes from the solution to Schrödinger's equation. Quantization of electron energies is a requirement in order to solve the equation. This is unlike the Bohr model, in which quantization was simply assumed with no mathematical basis.

Recall that in the Bohr model, the exact path of the electron was restricted to very well-defined circular orbits around the nucleus. The quantum mechanical model is a radical departure from that. Solutions to the Schrödinger wave equation, called wave functions, give only the probability of finding an electron at a given point around the nucleus. Electrons do not travel around the nucleus in simple circular orbits.

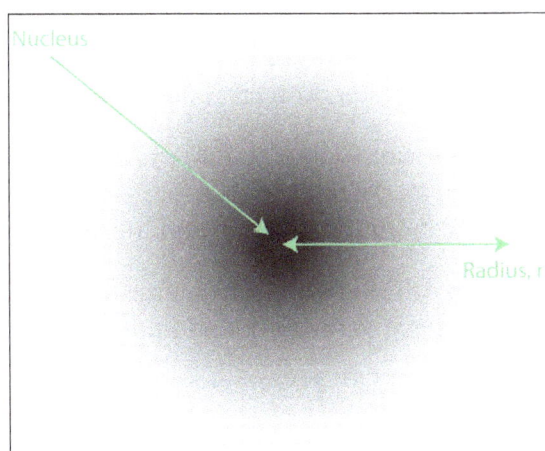

An electron cloud: The darker region nearer the nucleus indicates a high probability of finding the electron, while the lighter region further from the nucleus indicates a lower probability of finding the electron.

The location of the electrons in the quantum mechanical model of the atom is often referred to as an electron cloud. The electron cloud can be thought of in the following way: Imagine placing a square piece of paper on the floor with a dot in the circle representing the nucleus. Now take a marker and drop it onto the paper repeatedly, making small marks at each point the marker hits.

If you drop the marker many, many times, the overall pattern of dots will be roughly circular. If you aim toward the center reasonably well, there will be more dots near the nucleus and progressively fewer dots as you move away from it. Each dot represents a location where the electron could be at any given moment. Because of the uncertainty principle, there is no way to know exactly where the electron is. An electron cloud has variable densities: a high density where the electron is most likely to be and a low density where the electron is least likely to be.

In order to specifically define the shape of the cloud, it is customary to refer to the region of space within which there is a 90 probability of finding the electron. This is called an orbital, the three-dimensional region of space that indicates where there is a high probability of finding an electron.

ATOMIC ORBITALS

Orbitals and Orbits

When a planet moves around the sun, you can plot a definite path for it which is called an orbit. A simple view of the atom looks similar and you may have pictured the electrons as orbiting around the nucleus. The truth is different, and electrons in fact inhabit regions of space known as orbitals.

Orbits and orbitals sound similar, but they have quite different meanings. It is essential that you understand the difference between them.

The Impossibility of Drawing Orbits for Electrons

To plot a path for something you need to know exactly where the object is and be able to work out exactly where it's going to be an instant later. You can't do this for electrons.

The Heisenberg Uncertainty Principle says - loosely - that you can't know with certainty both where an electron is and where it's going next. (What it actually says is that it is impossible to define with absolute precision, at the same time, both the position and the momentum of an electron.)

That makes it impossible to plot an orbit for an electron around a nucleus. Is this a big problem? No. If something is impossible, you have to accept it and find a way around it.

Hydrogen's Electron - The 1s Orbital

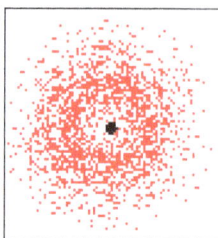

A 1s orbital.

Suppose you had a single hydrogen atom and at a particular instant plotted the position of the one electron. Soon afterwards, you do the same thing, and find that it is in a new position. You have no idea how it got from the first place to the second.

You keep on doing this over and over again, and gradually build up a sort of 3D map of the places that the electron is likely to be found.

In the hydrogen case, the electron can be found anywhere within a spherical space surrounding the nucleus. The diagram shows a cross-section through this spherical space.

95% of the time (or any other percentage you choose), the electron will be found within a fairly easily defined region of space quite close to the nucleus. Such a region of space is called an orbital. You can think of an orbital as being the region of space in which the electron lives.

What is the electron doing in the orbital? We don't know, we can't know, and so we just ignore the problem. All you can say is that if an electron is in a particular orbital it will have a particular definable energy.

Each orbital has a name.

The orbital occupied by the hydrogen electron is called a 1s orbital. The "1" represents the fact that the orbital is in the energy level closest to the nucleus. The "s" tells you about the shape of the orbital. s orbitals are spherically symmetric around the nucleus - in each case, like a hollow ball made of rather chunky material with the nucleus at its centre.

The orbital on the left is a 2s orbital. This is similar to a 1s orbital except that the region where there is the greatest chance of finding the electron is further from the nucleus - this is an orbital at the second energy level.

If you look carefully, you will notice that there is another region of slightly higher electron density (where the dots are thicker) nearer the nucleus. ("Electron density" is another way of talking about how likely you are to find an electron at a particular place.)

2s (and 3s, 4s, etc) electrons spend some of their time closer to the nucleus than you might expect. The effect of this is to slightly reduce the energy of electrons in s orbitals. The nearer the nucleus the electrons get, the lower their energy.

3s, 4s (etc) orbitals get progressively further from the nucleus.

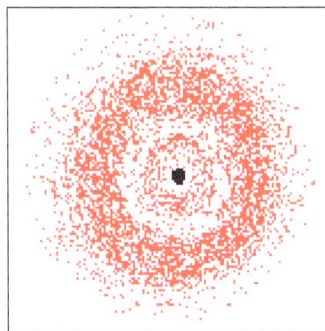

A 2s orbital.

p Orbitals

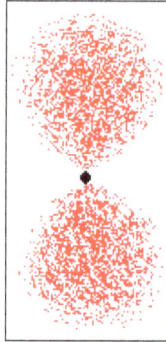

A p orbital.

Not all electrons inhabit s orbitals (in fact, very few electrons live in s orbitals). At the first energy level, the only orbital available to electrons is the 1s orbital, but at the second level, as well as a 2s orbital, there are also orbitals called 2p orbitals.

A p orbital is rather like 2 identical balloons tied together at the nucleus. The diagram on the left is a cross-section through that 3-dimensional region of space. Once again, the orbital shows where there is a 95% chance of finding a particular electron.

Unlike an s orbital, a p orbital points in a particular direction - the one drawn points up and down the page.

At any one energy level it is possible to have three absolutely equivalent p orbitals pointing mutually at right angles to each other. These are arbitrarily given the symbols p_x, p_y and p_z. This is simply for convenience - what you might think of as the x, y or z direction changes constantly as the atom tumbles in space.

The p orbitals at the second energy level are called 2px, 2py and 2pz. There are similar orbitals at subsequent levels - 3px, 3py, 3pz, 4px, 4py, 4pz and so on.

All levels except for the first level have p orbitals. At the higher levels the lobes get more elongated, with the most likely place to find the electron more distant from the nucleus.

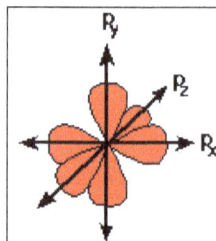

d and f Orbitals

In addition to s and p orbitals, there are two other sets of orbitals which become available for electrons to inhabit at higher energy levels. At the third level, there is a set of five d orbitals (with complicated shapes and names) as well as the 3s and 3p orbitals $\left(3p_x, 3p_y, 3p_z\right)$. At the third level there are a total of nine orbitals altogether.

At the fourth level, as well the 4s and 4p and 4d orbitals there are an additional seven f orbitals - 16 orbitals in all. s, p, d and f orbitals are then available at all higher energy levels as well.

For the moment, you need to be aware that there are sets of five d orbitals at levels from the third level upwards, but you probably won't be expected to draw them or name them. Apart from a passing reference, you won't come across f orbitals at all.

Fitting Electrons into Orbitals

You can think of an atom as a very bizarre house (like an inverted pyramid!) - with the nucleus living on the ground floor, and then various rooms (orbitals) on the higher floors occupied by the electrons. On the first floor there is only 1 room (the 1s orbital); on the second floor there are 4 rooms (the 2s, 2px, 2py and 2pz orbitals); on the third floor there are 9 rooms (one 3s orbital, three 3p orbitals and five 3d orbitals); and so on. But the rooms aren't very big; each orbital can only hold 2 electrons.

A convenient way of showing the orbitals that the electrons live in is to draw "electrons-in-boxes".

"Electrons-in-boxes"

Orbitals can be represented as boxes with the electrons in them shown as arrows. Often an up-arrow and a down-arrow are used to show that the electrons are in some way different.

A 1s orbital holding 2 electrons would be drawn as shown on the right, but it can be written even more quickly as 1s2. This is read as "one s two" - not as "one s squared".

1s.

You mustn't confuse the two numbers in this notation:

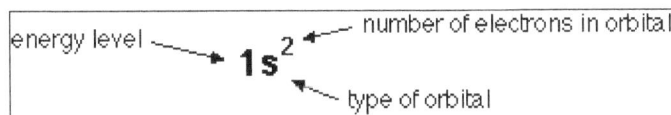

The Order of Filling Orbitals - the Aufbau Principle

Aufbau is a German word meaning building up or construction. We imagine that as you go from one atom to the next in the Periodic Table, you can work out the electronic structure of the next atom by fitting an extra electron into the next available orbital.

Electrons fill low energy orbitals (closer to the nucleus) before they fill higher energy ones. Where there is a choice between orbitals of equal energy, they fill the orbitals singly as far as possible.

This filling of orbitals singly where possible is known as Hund's rule. It only applies where the orbitals have exactly the same energies (as with p orbitals, for example), and helps to minimise the repulsions between electrons and so makes the atom more stable.

The diagram (not to scale) summarises the energies of the orbitals up to the 4p level that you will need to know when you are using the Aufbau Principle.

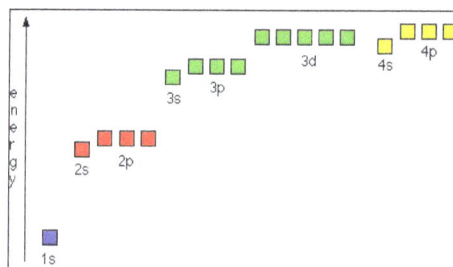

Notice that the s orbital always has a slightly lower energy than the p orbitals at the same energy level, so the s orbital always fills with electrons before the corresponding p orbitals.

The real oddity is the position of the 3d orbitals. They are at a slightly higher level than the 4s - and so it is the 4s orbital which you fill first, followed by all the 3d orbitals and then the 4p orbitals.

Similar confusion occurs at higher levels, with so much overlap between the energy levels that you don't fill the 4f orbitals until after the 6s, for example.

For UK-based exam purposes, you simply have to remember that when you are using the Aufbau Principle, you fill the 4s orbital before the 3d orbitals. The same thing happens at the next level as well - you fill the 5s orbital before the 4d orbitals. All the other complications are beyond the scope of this site.

Knowing the order of filling is central to understanding how to write electronic structures.

QUANTUM NUMBERS

Quantum numbers describe values of conserved quantities in the dynamics of a quantum system. In the case of electrons, the quantum numbers can be defined as "the sets of numerical values which give acceptable solutions to the Schrödinger wave equation for the hydrogen atom". An important aspect of quantum mechanics is the quantization of the observable quantities, since quantum numbers are discrete sets of integers or half-integers, although they could approach infinity in some cases. This distinguishes quantum mechanics from classical mechanics where the values that characterize the system such as mass, charge, or momentum, range continuously. Quantum numbers often describe specifically the energy levels of electrons in atoms, but other possibilities include angular momentum, spin, etc. An important family is flavour quantum numbers – internal quantum numbers which determine the type of a particle and its interactions with other particles through the forces. Any quantum system can have one or more quantum numbers; it is thus difficult to list all possible quantum numbers.

How Many Quantum Numbers Exist?

The question of how many quantum numbers are needed to describe any given system has no universal answer. Hence for each system one must find the answer for a full analysis of the system. A quantized system requires at least one quantum number. The dynamics of any quantum system are described by a quantum Hamiltonian, H. There is one quantum number of the system corresponding to the energy, that is to say the eigenvalue of the Hamiltonian. There is also one quantum number for each operator O that commutes with the Hamiltonian. These are all the quantum numbers that the system can have. Note that the operators O defining the quantum numbers should be independent of each other. Often, there is more than one way to choose a set of independent operators. Consequently, in different situations different sets of quantum numbers may be used for the description of the same system.

Electron in an Atom

Four quantum numbers can describe an electron in an atom completely:

- Principal quantum number (n).

- Azimuthal quantum number (ℓ).

- Magnetic quantum number (m).

- Spin quantum number (s).

The spin-orbital interaction, however, relates these numbers. Thus, a complete description of the system can be given with fewer quantum numbers, if orthogonal choices are made for these basis vectors.

Specificity

It is important to specify the electron being referred to. This may, for example, be the highest occupied orbital electron; the actual differentiating electron; or the differentiating electron according to the aufbau approximation. In lanthanum, for example, the electrons involved are in the 6s; 5d; and 4f orbitals, respectively. In this case the principle quantum numbers are 6, 5, and 4. The "differentiating electron" means the electron that differentiates an element from the previous one.

Common Terminology

The model used here describes electrons using four quantum numbers, n, ℓ, $m\ell$, ms, given below. It is also the common nomenclature in the classical description of nuclear particle states (e.g. protons and neutrons). A quantum description of molecular orbitals require different quantum numbers, because the Hamiltonian and its symmetries are quite different.

Principal Quantum Number

This describes the electron shell, or energy level, of an electron. The value of n ranges from 1 to the shell containing the outermost electron of that atom, that is:

$$n = 1, \ 2, \ ...$$

For example, in caesium (Cs), the outermost valence electron is in the shell with energy level 6, so an electron in caesium can have an n value from 1 to 6.

For particles in a time-independent potential, it also labels the nth eigenvalue of Hamiltonian (H), that is the energy, E with the contribution due to angular momentum (the term involving J^2) left out. This number therefore has a dependence only on the distance between the electron and the nucleus (i.e., the radial coordinate, r). The average distance increases with n, and hence quantum states with different principal quantum numbers are said to belong to different shells.

Azimuthal Quantum Number

Also known as the angular quantum number or orbital quantum number), this describes the subshell, and gives the magnitude of the orbital angular momentum through the relation:

$$L^2 = \hbar^2 \ell(\ell + 1)$$

In chemistry and spectroscopy, $\ell = 0$ is called an s orbital, $\ell = 1$ a p orbital, $\ell = 2$ a d orbital, and $\ell = 3$ an f orbital.

The value of ℓ ranges from o to $n - 1$, so the first p orbital $(\ell = 1)$ appears in the second electron shell $(n = 2)$, the first d orbital $(\ell = 2)$ appears in the third shell $(n = 3)$, and so on:

$$\ell = 0, 1, 2, ..., n - 1$$

A quantum number beginning in $n = 3, \ell = 0$, describes an electron in the s orbital of the third electron shell of an atom. In chemistry, this quantum number is very important, since it specifies the shape of an atomic orbital and strongly influences chemical bonds and bond angles.

Magnetic Quantum Number

This describes the specific orbital (or "cloud") within that subshell, and yields the projection of the orbital angular momentum along a specified axis:

$$L_z = m_\ell \hbar$$

The values of m_ℓ range from $-\ell$ to ℓ, with integer intervals:

The s subshell $(\ell = 0)$ contains only one orbital, and therefore the mℓ of an electron in an s orbital will always be o. The p subshell $(\ell = 1)$ contains three orbitals (in some systems, depicted as three "dumbbell-shaped" clouds), so the mℓ of an electron in a p orbital will be –1, 0, or 1. The d subshell $(\ell = 2)$ contains five orbitals, with mℓ values of –2, –1, 0, 1, and 2.

Spin Quantum Number

This describes the spin (intrinsic angular momentum) of the electron within that orbital, and gives the projection of the spin angular momentum S along the specified axis:

$$S_z = m_s \hbar.$$

In general, the values of ms range from –s to s, where s is the spin quantum number, an intrinsic property of particles:

$$m_s = -s, -s + 1, -s + 2, ..., s - 2, s - 1, s.$$

An electron has spin number $s = \frac{1}{2}$, consequently ms will be $\pm\frac{1}{2}$, referring to "spin up" and "spin down" states. Each electron in any individual orbital must have different quantum numbers because of the Pauli exclusion principle, therefore an orbital never contains more than two electrons.

Rules

There are no universal fixed value for m_ℓ and m_s values. Rather, the m_ℓ and m_s values are arbitrary. The only requirement is that the naming schematic used within a particular set of calculations or descriptions must be consistent (e.g. the orbital occupied by the first electron in a p orbital could be described as $m_\ell = -1$ or $m_\ell = 0$ or $m_\ell = 1$, but the mℓ value of the next unpaired electron in that orbital must be different; yet, the mℓ assigned to electrons in other orbitals again can be $m_\ell = -1$ or $m_\ell = 0$ or $m_\ell = 1$).

These rules are summarized as follows:

Name	Symbol	Orbital meaning	Range of values	Value examples
principal quantum number	n	shell	$1 \leq$	$n = 1, 2, 3, \ldots$
azimuthal quantum number (angular momentum)	ℓ	subshell (s orbital is listed as 0, p orbital as 1 etc.)	$0 \leq \ell \leq n-1$	*for* $n = 3$: $\ell = 0, 1, 2 (s, p, d)$
magnetic quantum number (projection of angular momentum)	m_ℓ	energy shift (orientation of the subshell's shape)	$-\ell \leq m_\ell \leq \ell$	*for* $\ell = 2$: $m_\ell = -2, -1, 0, 1, 2$
spin projection quantum number	m_s	spin of the electron $\left(-\frac{1}{2}\right) =$ "spin down", $\frac{1}{2} =$ "spin up")	$-s \leq m_s \leq s$	*for an electron* $s = \frac{1}{2}$, *so* $m_s = -\frac{1}{2}, +\frac{1}{2}$

Example: The quantum numbers used to refer to the outermost valence electrons of a carbon (C) atom, which are located in the 2p atomic orbital, are; $n = 2$ (2nd electron shell), $\ell = 1$ (p orbital subshell), $m\ell = 1, 0, -1$, $ms = \frac{1}{2}$ (parallel spins).

Results from spectroscopy indicated that up to two electrons can occupy a single orbital. However two electrons can never have the same exact quantum state nor the same set of quantum numbers according to Hund's rules, which addresses the Pauli exclusion principle. A fourth quantum number with two possible values was added as an ad hoc assumption to resolve the conflict; this supposition could later be explained in detail by relativistic quantum mechanics and from the results of the renowned Stern–Gerlach experiment.

Total Angular Momenta Numbers

Total Momentum of a Particle

When one takes the spin–orbit interaction into consideration, the L and S operators no longer commute with the Hamiltonian, and their eigenvalues therefore change over time. Thus another set of quantum numbers should be used. This set includes;

- The total angular momentum quantum number,

$$j = |\ell \pm s|,$$

  which gives the total angular momentum through the relation,

$$J^2 = \hbar^2 j(j + 1).$$

- The projection of the total angular momentum along a specified axis,

$$m_j = -j, \; -j + 1, \; -j + 2, \; ..., j - 2, j - 1, j,$$

  analogous to the above and satisfies,

$$m_j = m_\ell + m_s \, and \left| m_\ell + m_s \right| \leq j.$$

- Parity: This is the eigenvalue under reflection: positive (+1) for states which came from even ℓ and negative (−1) for states which came from odd ℓ. The former is also known as even parity and the latter as odd parity, and is given by:

$$P = \left(-1\right)^\ell.$$

For example, consider the following 8 states, defined by their quantum numbers:

	n	ℓ	m_ℓ	m_s
(1)	2	1	1	$+\dfrac{1}{2}$
(2)	2	1	1	$-\dfrac{1}{2}$
(3)	2	2	0	$+\dfrac{1}{2}$
(4)	2	1	0	$-\dfrac{1}{2}$
(5)	2	1	-1	$+\dfrac{1}{2}$
(6)	2	1	-1	$-\dfrac{1}{2}$

$\ell + s$	$\ell - s$	$m_\ell + m_s$
$\dfrac{3}{2}$	$\dfrac{1}{2}$	$\dfrac{3}{2}$
$\dfrac{3}{2}$	$\dfrac{1}{2}$	$\dfrac{1}{2}$
$\dfrac{3}{2}$	$\dfrac{1}{2}$	$\dfrac{1}{2}$
$\dfrac{3}{2}$	$\dfrac{1}{2}$	$-\dfrac{1}{2}$
$\dfrac{3}{2}$	$\dfrac{1}{2}$	$-\dfrac{1}{2}$
$\dfrac{1}{2}$	$\dfrac{1}{2}$	—

(7)	2	0	0	$+\dfrac{1}{2}$
(8)	2	0	0	$-\dfrac{1}{2}$

$\dfrac{1}{2}$	$-\dfrac{1}{2}$	$\dfrac{1}{2}$
$\dfrac{1}{2}$	$-\dfrac{1}{2}$	$-\dfrac{1}{2}$

The quantum states in the system can be described as linear combination of these 8 states. However, in the presence of spin–orbit interaction, if one wants to describe the same system by 8 states that are eigenvectors of the Hamiltonian (i.e. each represents a state that does not mix with others over time), we should consider the following 8 states:

j	m_j	parity	
$\dfrac{3}{2}$	$\dfrac{3}{2}$	odd	coming from state (1) above
$\dfrac{3}{2}$	$\dfrac{1}{2}$	odd	coming from states (2) and (3) above
$\dfrac{3}{2}$	$-\dfrac{1}{2}$	odd	coming from states (4) and (5) above
$\dfrac{3}{2}$	$-\dfrac{3}{2}$	odd	coming from state (6) above
$\dfrac{1}{2}$	$\dfrac{1}{2}$	odd	coming from states (2) and (3) above
$\dfrac{1}{2}$	$-\dfrac{1}{2}$	odd	coming from states (4) and (5) above
$\dfrac{1}{2}$	$\dfrac{1}{2}$	even	coming from state (7) above
$\dfrac{1}{2}$	$-\dfrac{1}{2}$	even	coming from state (8) above

Nuclear Angular Momentum Quantum Numbers

In nuclei, the entire assembly of protons and neutrons (nucleons) has a resultant angular momentum due to the angular momenta of each nucleon, usually denoted I. If the total angular momentum of a neutron is $j_n = \ell + s$ and for a proton is $j_p = \ell + s$ (where s for protons and neutrons happens to be again) then the nuclear angular momentum quantum numbers I are given by:

$$I = |j_n - j_p|, \; |j_n - j_p| + 1, \; |j_n - j_p| + 2, \; ..., \; (j_n + j_p) - 2, \; (j_n + j_p) - 1, \; (j_n + j_p)$$

Parity with the number I is used to label nuclear angular momentum states, examples for some isotopes of hydrogen (H), carbon (C), and sodium (Na) are;

$${}^{1}_{1}\text{H}\, I = \left(\frac{1}{2}\right)^{+} \quad {}^{9}_{6}C\, I = \left(\frac{3}{2}\right)^{-} \quad {}^{20}_{11}\text{Na}\, I = 2^{+}$$

$${}^{2}_{1}H\, I = 1^{+} \quad {}^{10}_{6}C\, I = 0^{+} \quad {}^{21}_{11}\text{Na}\, I = \left(\frac{3}{2}\right)^{+}$$

$${}^{3}_{1}H\, I = \left(\frac{1}{2}\right)^{+} \quad {}^{11}_{6}C\, I = \left(\frac{3}{2}\right)^{-} \quad {}^{22}_{11}\text{Na}\, I = 3^{+}$$

$${}^{12}_{6}C\, I = 0^{+} \quad {}^{23}_{11}\text{Na}\, I = \left(\frac{3}{2}\right)^{+}$$

$${}^{13}_{6}C\, I = \left(\frac{1}{2}\right)^{-} \quad {}^{24}_{11}\text{Na}\, I = 4^{+}$$

$${}^{14}_{6}C\, I = 0^{+} \quad {}^{25}_{11}\text{Na}\, I = \left(\frac{5}{2}\right)^{+}$$

$${}^{15}_{6}C\, I = \left(\frac{1}{2}\right)^{+} \quad {}^{26}_{11}\text{Na}\, I = 3^{+}$$

The reason for the unusual fluctuations in I, even by differences of just one nucleon, are due to the odd and even numbers of protons and neutrons – pairs of nucleons have a total angular momentum of zero (just like electrons in orbitals), leaving an odd or even number of unpaired nucleons. The property of nuclear spin is an important factor for the operation of NMR spectroscopy in organic chemistry, and MRI in nuclear medicine, due to the nuclear magnetic moment interacting with an external magnetic field.

Elementary Particles

Elementary particles contain many quantum numbers which are usually said to be intrinsic to them. However, it should be understood that the elementary particles are quantum states of the standard model of particle physics, and hence the quantum numbers of these particles bear the same relation to the Hamiltonian of this model as the quantum numbers of the Bohr atom does to its Hamiltonian. In other words, each quantum number denotes a symmetry of the problem. It is more useful in quantum field theory to distinguish between spacetime and internal symmetries.

Typical quantum numbers related to spacetime symmetries are spin (related to rotational symmetry), the parity, C-parity and T-parity (related to the Poincaré symmetry of spacetime). Typical internal symmetriesare lepton number and baryon number or the electric charge.

Multiplicative Quantum Numbers

A minor but often confusing point is as follows: most conserved quantum numbers are additive, so in an elementary particle reaction, the sum of the quantum numbers should be the same before and after the reaction. However, some, usually called a parity, are multiplicative; i.e., their product is conserved. All multiplicative quantum numbers belong to a symmetry (like parity) in which applying the symmetry transformation twice is equivalent to doing nothing (involution).

AUFBAU PRINCIPLE

The aufbau principle (also called the building-up principle or the aufbau rule) states that in the ground state of an atom or ion, electrons fill atomic orbitals of the lowest available energy levels before occupying higher levels. For example, the 1s shell is filled before the 2s subshell is occupied. In this way, the electrons of an atom or ion form the most stable electron configuration possible. An example is the configuration $1s^2\ 2s^2\ 2p^6\ 3s^2\ 3p^3$ for the phosphorus atom, meaning that the 1s subshell has 2 electrons, and so on.

Electron behavior is elaborated by other principles of atomic physics, such as Hund's rule and the Pauli exclusion principle. Hund's rule asserts that if multiple orbitals of the same energy are available, electrons will occupy different orbitals singly before any are occupied doubly. If double occupation does occur, the Pauli exclusion principle requires that electrons which occupy the same orbital must have different spins (+1/2 and −1/2).

As we pass from one element to another of next higher atomic number, one proton and one electron are added each time to the neutral atom. The maximum number of electrons in any shell is $2n^2$, where n is the principal quantum number. The maximum number of electrons in a subshell (s, p, d or f) is equal to $2(2\ell+1)$ where $\ell = 0, 1, 2, 3....$. Thus these subshells can have a maximum of 2, 6, 10 and 14 electrons respectively. In the ground state the electronic configuration can be built up by placing electrons in the lowest available orbitals until the total number of electrons added is equal to the atomic number. Thus orbitals are filled in the order of increasing energy, using two general rules to help predict electronic configurations:

- Electrons are assigned to orbitals in order of increasing value of $(n+\ell)$.

- For subshells with the same value of $(n+\ell)$, electrons are assigned first to the sub shell with lower n.

A version of the aufbau principle known as the nuclear shell model is used to predict the configuration of protons and neutrons in an atomic nucleus.

Madelung Energy Ordering Rule

In neutral atoms, the approximate order in which subshells are filled is given by the $n + \ell$ rule, also known as the:

- The Madelung rule (after Erwin Madelung);

- The Janet rule (after Charles Janet);

- The Klechkowsky rule (after Vsevolod Klechkovsky);

- The aufbau approximation;

- The Uncle Wiggly path; or

- The diagonal rule.

Here n represents the principal quantum number and ℓ the azimuthal quantum number; the values ℓ = 0, 1, 2, 3 correspond to the s, p, d, and f labels, respectively. The subshell ordering by this rule is 1s, 2s, 2p, 3s, 3p, 4s, 3d, 4p, 5s, 4d, 5p, 6s, 4f, 5d, 6p, 7s, 5f, 6d, 7p, 8s, For example titanium (Z = 22) has the ground-state configuration $1s^2\ 2s^2\ 2p^6\ 3s^2\ 3p^6\ 4s^2\ 3d^2$.

Other authors write the orbitals always in order of increasing n, such as Ti (Z = 22) $1s^2\ 2s^2\ 2p^6\ 3s^2\ 3p^6\ 3d^2\ 4s^2$. This can be called "leaving order", since if the atom is ionized, electrons leave in the order 4s, 3d, 3p, 3s, etc. For a given neutral atom, the two notations are equivalent since only the orbital occupancies have physical significance.

Orbitals with a lower $n + \ell$ value are filled before those with higher $n + \ell$ values. In the case of equal $n + \ell$ values, the orbital with a lower n value is filled first. The Madelung energy ordering rule applies only to neutral atoms in their ground state. There are ten elements among the transition metals and ten elements among the lanthanides and actinides for which the Madelung rule predicts an electron configuration that differs from that determined experimentally. These exceptions can be viewed as turbulence along a flight path; once passed the flight path returns to normal.

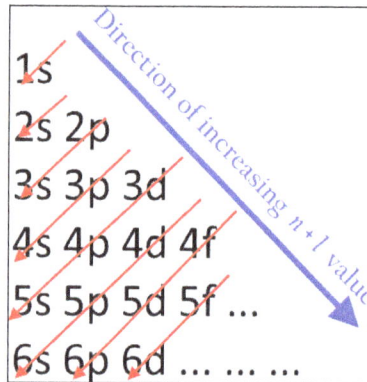

The states crossed by same red arrow have same $n + \ell$ value.
The direction of the red arrow indicates the order of state filling.

Exceptions to the Rule in the Transition Metals

The valence d-subshell "borrows" one electron (in the case of palladium two electrons) from the valence s-subshell.

Atom	$_{24}$Cr	$_{29}$Cu	$_{41}$Nb	$_{42}$Mo	$_{44}$Ru	$_{45}$Rh	$_{46}$Pd	$_{47}$Ag	$_{78}$Pt	$_{79}$Au
Core electrons	[Ar]	[Ar]	[Kr]	[Kr]	[Kr]	[Kr]	[Kr]	[Kr]	[Xe]	[Xe]
Madelung Rule	$3d^4 4s^2$	$3d^9 4s^2$	$4d^3 5s^2$	$4d^4 5s^2$	$4d^6 5s^2$	$4d^7 5s^2$	$4d^8 5s^2$	$4d^9 5s^2$	$4f^{14} 5d^8 6s^2$	$4f^{14} 5d^9 6s^2$

Experi-ment	$3d^5 4s^1$	$3d^{10} 4s^1$	$4d^4 5s^1$	$4d^5 5s^1$	$4d^7 5s^1$	$4d^8 5s^1$	$4d^{10}$	$4d^{10} 5s^1$	$4f^{14} 5d^9 6s^1$	$4f^{14} 5d^{10} 6s^1$

For example, in copper $_{29}$Cu , according to the Madelung rule, the 4s orbital $(n + \ell = 4 + 0 = 4)$ is occupied before the 3d orbital $(n + \ell = 3 + 2 = 5)$. The rule then predicts the electron configuration $1s^2 2s^2 2p^6 3s^2 3p^6 3d^9 4s^2$, abbreviated [Ar] $3d^9 4s^2$ where [Ar] denotes the configuration of the preceding noble gas argon. However, the measured electron configuration of the copper atom is [Ar] $3d^{10} 4s^1$. By filling the 3d orbital, copper can be in a lower energy state.

Exceptions among the Lanthanides and Actinides

The valence d-subshell typically "borrows" one electron (in the case of thorium two electrons) from the valence f-subshell. For example, in uranium $_{92}$U , according to the Madelung rule, the 5f orbital $(n + \ell = 5 + 3 = 8)$ is occupied before the 6d orbital $(n + \ell = 6 + 2 = 8)$. The rule then predicts the electron configuration [Rn] $5f^4 7s^2$ where [Rn] denotes the configuration of the preceding noble gas radon. However, the measured electron configuration of the uranium atom is [Rn] $5f^3 6d^1 7s^2$.

A special exception is lawrencium $_{103}$Lr , where the 6d electron predicted by the Madelung rule is replaced by a 7p electron: the rule predicts [Rn] $5f^{14} 6d^1 7s^2$, but the measured configuration is [Rn] $5f^{14} 7s^2 7p^1$.

Atom	$_{57}$La	$_{58}$Ce	$_{64}$Gd	$_{89}$Ac	$_{90}$Th	$_{91}$Pa	$_{92}$U	$_{93}$Np	$_{96}$Cm	$_{103}$Lr
Core electrons	[Xe]	[Xe]	[Xe]	[Rn]	[Rn]	[Rn]	[Rn]	[Rn]	[Rn]	[Rn]
Madelung Rule	$4f^1 6s^2$	$4f^2 6s^2$	$4f^8 6s^2$	$5f^1 7s^2$	$5f^2 7s^2$	$5f^3 7s^2$	$5f^4 7s^2$	$5f^5 7s^2$	$5f^8 7s^2$	$5f^{14} 6d^1 7s^2$
Experiment	$5d^1 6s^2$	$4f^1 5d^1 6s^2$	$4f^7 5d^1 6s^2$	$6d^1 7s^2$	$6d^2 7s^2$	$5f^2 6d^1 7s^2$	$5f^3 6d^1 7s^2$	$5f^4 6d^1 7s^2$	$5f^7 6d^1 7s^2$	$5f^{14} 7s^2 7p^1$

History

The Aufbau Principle in the New Quantum Theory

The principle takes its name from the German, Aufbauprinzip, "building-up principle", rather than being named for a scientist. It was formulated by Niels Bohr and Wolfgang Pauli in the early 1920s. This was an early application of quantum mechanics to the properties of electrons, and explained chemical properties in physical terms. Each added electron is subject to the electric field created by the positive charge of the atomic nucleus and the negative charge of other electrons that are bound to the nucleus. Although in hydrogen there is no energy difference between orbitals with the same principal quantum number n, this is not true for the outer electrons of other atoms.

In the old quantum theory prior to quantum mechanics, electrons were supposed to occupy classical elliptical orbits. The orbits with the highest angular momentum are 'circular orbits' outside the inner electrons, but orbits with low angular momentum (s- and p-orbitals) have high orbital eccentricity, so that they get closer to the nucleus and feel on average a less strongly screened nuclear charge.

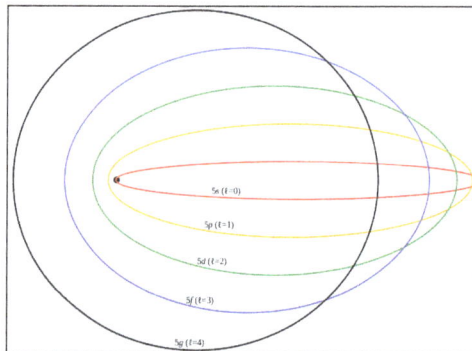

In the old quantum theory, orbits with low angular momentum (s- and p-orbitals) get closer to the nucleus.

The $n + \ell$ Energy Ordering Rule

A periodic table in which each row corresponds to one value of $n + \ell$ was suggested by Charles Janet in 1928, and in 1930 he made explicit the quantum basis of this pattern, based on knowledge of atomic ground states determined by the analysis of atomic spectra. This table came to be referred to as the left-step table. Janet "adjusted" some of the actual n + l values of the elements, since they did not accord with his energy ordering rule, and he considered that the discrepancies involved must have arisen from measurement errors. In the event, the actual values were correct and the $n + \ell$ energy ordering rule turned out to be an approximation rather than a perfect fit.

In 1936, the German physicist Erwin Madelung proposed this as an empirical rule for the order of filling atomic subshells, and most English-language sources therefore refer to the Madelung rule. Madelung may have been aware of this pattern as early as 1926. In 1962 the Russian agricultural chemist V.M. Klechkowski proposed the first theoretical explanation for the importance of the sum $n + \ell$ based on the statistical Thomas–Fermi model of the atom. Many French- and Russian-language sources therefore refer to the Klechkowski rule.

In recent years it has been noted that the order of filling orbitals in neutral atoms does not always correspond to the order of adding or removing electrons for a given atom. For example, in the fourth row of the periodic table the Madelung rule indicates that the 4s orbital is occupied before the 3d. The neutral atom ground state configurations are therefore K = (Ar)4s, Ca = (Ar)4s², Sc = (Ar)4s²3d, etc. However, if a scandium atom is ionized by removing electrons (only), the configurations are Sc = (Ar) 4s23d, Sc⁺ = (Ar) 4s3d, Sc²⁺ = (Ar)3d. The orbital energies and their order depend on the nuclear charge; 4s is lower than 3d as per the Madelung rule in K with 19 protons, but 3d is lower in Sc²⁺ with 21 protons. The Madelung rule should only be used for neutral atoms.

In addition to there being ample experimental evidence to support this view, it makes the explanation of the order of ionization of electrons in this and other transition metals more intelligible, given that 4s electrons are invariably preferentially ionized.

HUND'S RULES

In atomic physics, Hund's rules refers to a set of rules that German physicist Friedrich Hund formulated around 1927, which are used to determine the term symbol that corresponds to the ground state of a multi-electron atom. The first rule is especially important in chemistry, where it is often referred to simply as Hund's Rule.

The three rules are:

- For a given electron configuration, the term with maximum multiplicity has the lowest energy. The multiplicity is equal to $2S+1$, where S is the total spin angular momentum for all electrons. Therefore, the term with lowest energy is also the term with maximum S.

- For a given multiplicity, the term with the largest value of the total orbital angular momentum quantum number L has the lowest energy.

- For a given term, in an atom with outermost subshell half-filled or less, the level with the lowest value of the total angular momentum quantum number J (for the operator $J = L + S$) lies lowest in energy. If the outermost shell is more than half-filled, the level with the highest value of J is lowest in energy.

These rules specify in a simple way how usual energy interactions determine which term includes the ground state. The rules assume that the repulsion between the outer electrons is much greater than the spin–orbit interaction, which is in turn stronger than any other remaining interactions. This is referred to as the LS coupling regime.

Full shells and subshells do not contribute to the quantum numbers for total S, the total spin angular momentum and for L, the total orbital angular momentum. It can be shown that for full orbitals and suborbitals both the residual electrostatic energy (repulsion between electrons) and the spin–orbit interaction can only shift all the energy levels together. Thus when determining the ordering of energy levels in general only the outer valence electrons must be considered.

Rule 1

Due to the Pauli exclusion principle, two electrons cannot share the same set of quantum numbers within the same system; therefore, there is room for only two electrons in each spatial orbital. One of these electrons must have, (for some chosen direction z) $m_s = \frac{1}{2}$, and the other must have $m_s = -\frac{1}{2}$. Hund's first rule states that the lowest energy atomic state is the one that maximizes the total spin quantum number for the electrons in the open subshell. The orbitals of the subshell are each occupied singly with electrons of parallel spin before double occupation occurs. (This is occasionally called the "bus seat rule" since it is analogous to the behaviour of bus passengers who tend to occupy all double seats singly before double occupation occurs.)

Two different physical explanations have been given for the increased stability of high multiplicity states. In the early days of quantum mechanics, it was proposed that electrons in different orbitals are further apart, so that electron–electron repulsion energy is reduced. However, accurate quantum-mechanical calculations (starting in the 1970s) have shown that the reason is that the electrons in singly occupied orbitals are less effectively screened or shielded from the nucleus, so that such orbitals contract and electron–nucleus attraction energy becomes greater in magnitude (or decreases algebraically).

As an example, consider the ground state of silicon. The electronic configuration of Si is $1s^2\ 2s^2\ 2p^6\ 3s^2\ 3p^2$. We need to consider only the outer $3p^2$ electrons, for which it can be shown that the possible terms allowed by the Pauli exclusion principle are ^1D, ^3P, and ^1S. Hund's first rule now states that the ground state term is ^3P (triplet P), which has S = 1. The superscript 3 is the value of the multiplicity = $2S + 1 = 3$. The diagram shows the state of this term with M_L = 1 and M_S = 1.

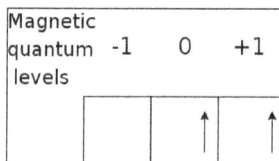

Hund's rules applied to Si. The up arrows signify electrons with up-spin. The boxes represent different magnetic quantum numbers.

Rule 2

This rule deals with reducing the repulsion between electrons. It can be understood from the classical picture that if all electrons are orbiting in the same direction (higher orbital angular momentum) they meet less often than if some of them orbit in opposite directions. In the latter case the repulsive force increases, which separates electrons. This adds potential energy to them, so their energy level is higher.

Example:

For silicon there is only one triplet term, so the second rule is not required. The lightest atom that requires the second rule to determine the ground state term is titanium (Ti, Z = 22) with electron configuration $1s^2 2s^2 2p^6 3s^2 3p^6 3d^2 4s^2$. In this case the open shell is $3d^2$ and the allowed terms include three singlets ($^1S, ^1D$, and ^1G) and two triplets (^3P and ^3F). (Here the symbols S, P, D, F, and G indicate that the total orbital angular momentum quantum number has values 0, 1, 2, 3 and 4, respectively, analogous to the nomenclature for naming atomic orbitals.)

We deduce from Hund's first rule that the ground state term is one of the two triplets, and from Hund's second rule that this term is ^3F (with L = 3) rather than ^3P (with L = 1) There is no ^3G term since its ($M_L = 4, M_S = 1$) state would require two electrons each with ($M_L = 2, M_S = +1/2$), in violation of the Pauli principle. (Here $= M_L$ and M_S are the components of the total orbital angular momentum L and total spin S along the z-axis chosen as the direction of an external magnetic field.)

Rule 3

This rule considers the energy shifts due to spin–orbit coupling. In the case where the spin–orbit coupling is weak compared to the residual electrostatic interaction, L and S are still good quantum numbers and the splitting is given by:

$$\Delta E = \zeta(L,S)\{\mathbf{L}\cdot\mathbf{S}\}$$
$$= (1/2)\zeta(L,S)\{J(J+1)-L(L+1)-S(S+1)\}$$

The value of $\zeta(L,S)$ changes from plus to minus for shells greater than half full. This term gives the dependence of the ground state energy on the magnitude of J.

Examples:

The ^3P lowest energy term of Si consists of three levels, $J = 2,1,0$. With only two of six possible electrons in the shell, it is less than half-full and thus $^3P_0$ is the ground state. For sulfur (S) the lowest energy term is again ^3P with spin–orbit levels $J = 2,1,0$ but now there are four of six possible electrons in the shell so the ground state is $^3P_2$.

If the shell is half-filled then $L = 0$, and hence there is only one value of J (equal to S), which is the lowest energy state. For example, in phosphorus the lowest energy state has $S = 3/2$, $L = 0$ for three unpaired electrons in three 3p orbitals. Therefore, $J = S = 3/2$ and the ground state is $^4S_{3/2}$.

Excited States

Hund's rules work best for the determination of the ground state of an atom or molecule.

They are also fairly reliable (with occasional failures) for the determination of the lowest state of a given excited electronic configuration. Thus, in the helium atom, Hund's first rule correctly predicts that the 1s2s triplet state $\left(^3S\right)$ is lower than the 1s2s singlet $\left(^1S\right)$. Similarly for organic molecules, the same rule predicts that the first triplet state (denoted by T_1 in photochemistry) is lower than the first excited singlet state $\left(S_1\right)$, which is generally correct.

However Hund's rules should not be used to order states other than the lowest for a given configuration. For example, the titanium atom ground state configuration is $3d^2$ for which a naïve application of Hund's rules would suggest the ordering $^3F < ^3P < ^1G < ^1D < ^1S$. In reality, however, ^1D lies below ^1G.

ELECTRON CONFIGURATION

In atomic physics and quantum chemistry, the electron configuration is the distribution of electrons of an atom or molecule (or other physical structure) in atomic or molecular orbitals. For example, the electron configuration of the neon atom is $1s^2 2s^2 2p^6$, using the notation explained below.

Electronic configurations describe each electron as moving independently in an orbital, in an average field created by all other orbitals. Mathematically, configurations are described by Slater determinants or configuration state functions.

According to the laws of quantum mechanics, for systems with only one electron, a level of energy is associated with each electron configuration and in certain conditions, electrons are able to move from one configuration to another by the emission or absorption of a quantum of energy, in the form of a photon.

Knowledge of the electron configuration of different atoms is useful in understanding the structure of the periodic table of elements. This is also useful for describing the chemical bonds that hold atoms together. In bulk materials, this same idea helps explain the peculiar properties of lasers and semiconductors.

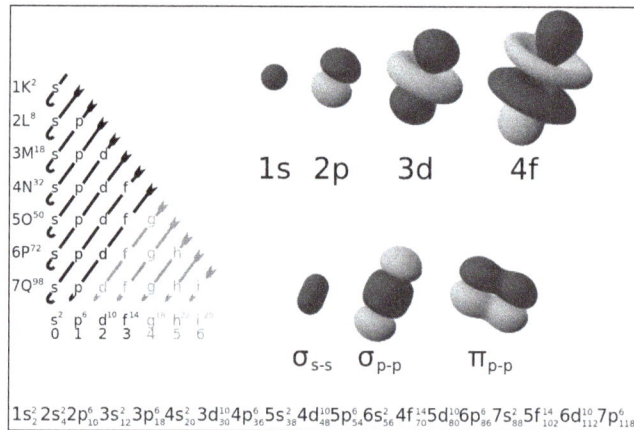

Electron atomic and molecular orbitals.

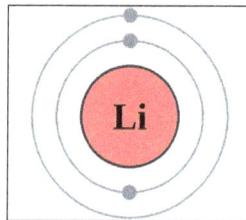

A Bohr diagram of lithium.

Shells and Subshells

Electron configuration was first conceived under the Bohr model of the atom, and it is still common to speak of shells and subshells despite the advances in understanding of the quantum-mechanical nature of electrons.

An electron shell is the set of allowed states that share the same principal quantum number, n (the number before the letter in the orbital label), that electrons may occupy. An atom's nth electron shell can accommodate $2n^2$ electrons, e.g. the first shell can accommodate 2 electrons, the second shell 8 electrons, the third shell 18 electrons and so on. The factor of two arises because the allowed states are doubled due to electron spin—each atomic orbital admits up to two otherwise identical electrons with opposite spin, one with a spin +1/2 (usually denoted by an up-arrow) and one with a spin –1/2 (with a down-arrow).

A subshell is the set of states defined by a common azimuthal quantum number, ℓ, within a shell. The values $\ell = 0, 1, 2, 3$ correspond to the s, p, d, and f labels, respectively. For example, the 3d subshell has $n = 3$ and $\ell = 2$. The maximum number of electrons that can be placed in a subshell is given by $2(2\ell+1)$. This gives two electrons in an s subshell, six electrons in a p subshell, ten electrons in a d subshell and fourteen electrons in an f subshell.

The numbers of electrons that can occupy each shell and each subshell arise from the equations of quantum mechanics, in particular the Pauli exclusion principle, which states that no two electrons in the same atom can have the same values of the four quantum numbers.

Notation

Physicists and chemists use a standard notation to indicate the electron configurations of atoms and molecules. For atoms, the notation consists of a sequence of atomic subshell labels (e.g. for phosphorus the sequence 1s, 2s, 2p, 3s, 3p) with the number of electrons assigned to each subshell placed as a superscript. For example, hydrogen has one electron in the s-orbital of the first shell, so its configuration is written $1s^1$. Lithium has two electrons in the 1s-subshell and one in the (higher-energy) 2s-subshell, so its configuration is written $1s^2\ 2s^1$ (pronounced "one-s-two, two-s-one"). Phosphorus (atomic number 15) is as follows: $1s^2\ 2s^2\ 2p^6\ 3s^2\ 3p^3$.

For atoms with many electrons, this notation can become lengthy and so an abbreviated notation is used. The electron configuration can be visualized as the core electrons, equivalent to the noble gas of the preceding period, and the valence electrons: each element in a period differs only by the last few subshells. Phosphorus, for instance, is in the third period. It differs from the second-period neon, whose configuration is $1s^2\ 2s^2\ 2p^6$, only by the presence of a third shell. The portion of its configuration that is equivalent to neon is abbreviated as [Ne], allowing the configuration of phosphorus to be written as [Ne] $3s^2\ 3p^3$ rather than writing out the details of the configuration of neon explicitly. This convention is useful as it is the electrons in the outermost shell that most determine the chemistry of the element.

For a given configuration, the order of writing the orbitals is not completely fixed since only the orbital occupancies have physical significance. For example, the electron configuration of the titanium ground state can be written as either [Ar] $4s^2\ 3d^2$ or $\left[Ar\right]\ 3d^2\ 4s^2$.. The first notation follows the order based on the Madelung rule for the configurations of neutral atoms; 4s is filled before 3d in the sequence Ar, K, Ca, Sc, Ti. The second notation groups all orbitals with the same value of n together, corresponding to the "spectroscopic" order of orbital energies that is the reverse of the order in which electrons are removed from a given atom to form positive ions; 3d is filled before 4s in the sequence Ti^{4+}, Ti^{3+}, Ti^{2+}, Ti^+, Ti.

The superscript 1 for a singly occupied subshell is not compulsory; for example aluminium may be written as either [Ne] $3s^2\ 3p^1$ or [Ne] $3s^2\ 3p$. It is quite common to see the letters of the orbital labels (s, p, d, f) written in an italic or slanting typeface, although the International Union of Pure and Applied Chemistry (IUPAC) recommends a normal typeface. The choice of letters originates from a now-obsolete system of categorizing spectral lines as "sharp", "principal", "diffuse" and "fundamental" (or "fine"), based on their observed fine structure: their modern usage indicates orbitals with an azimuthal quantum number, l, of 0, 1, 2 or 3 respectively. After "f", the sequence continues alphabetically "g", "h", "i"... (l = 4, 5, 6...), skipping "j", although orbitals of these types are rarely required.

The electron configurations of molecules are written in a similar way, except that molecular orbital labels are used instead of atomic orbital labels.

Energy — Ground State and Excited States

The energy associated to an electron is that of its orbital. The energy of a configuration is often approximated as the sum of the energy of each electron, neglecting the electron-electron interactions. The configuration that corresponds to the lowest electronic energy is called the ground state. Any other configuration is an excited state.

As an example, the ground state configuration of the sodium atom is $1s^2 2s^2 2p^6 3s^1$, as deduced from the Aufbau principle. The first excited state is obtained by promoting a 3s electron to the 3p orbital, to obtain the $1s^2 2s^2 2p^6 3p^1$ configuration, abbreviated as the 3p level. Atoms can move from one configuration to another by absorbing or emitting energy. In a sodium-vapor lamp for example, sodium atoms are excited to the 3p level by an electrical discharge, and return to the ground state by emitting yellow light of wavelength 589 nm.

Usually, the excitation of valence electrons (such as 3s for sodium) involves energies corresponding to photons of visible or ultraviolet light. The excitation of core electrons is possible, but requires much higher energies, generally corresponding to x-ray photons. This would be the case for example to excite a 2p electron of sodium to the 3s level and form the excited $1s^2 2s^2 2p^5 3s^2$ configuration.

The remainder of this topic deals only with the ground-state configuration, often referred to as "the" configuration of an atom or molecule.

Atoms: Aufbau Principle and Madelung Rule

The Aufbau principle (from the German Aufbau, "building up, construction") was an important part of Bohr's original concept of electron configuration. It may be stated as:

> a maximum of two electrons are put into orbitals in the order of increasing orbital energy: the lowest-energy orbitals are filled before electrons are placed in higher-energy orbitals.

The principle works very well (for the ground states of the atoms) for the first 18 elements, then decreasingly well for the following 100 elements. The modern form of the Aufbau principle describes an order of orbital energies given by Madelung's rule (or Klechkowski's rule). This rule was first stated by Charles Janet in 1929, rediscovered by Erwin Madelung in 1936, and later given a theoretical justification by V.M. Klechkowski:

- Orbitals are filled in the order of increasing $n + l$;

- Where two orbitals have the same value of $n + l$, they are filled in order of increasing n.

This gives the following order for filling the orbitals:

> 1s, 2s, 2p, 3s, 3p, 4s, 3d, 4p, 5s, 4d, 5p, 6s, 4f, 5d, 6p, 7s, 5f, 6d, 7p, (8s, 5g, 6f, 7d, 8p, and 9s)

In this list the orbitals in parentheses are not occupied in the ground state of the heaviest atom now known (Og, Z = 118).

The Aufbau principle can be applied, in a modified form, to the protons and neutrons in the atomic nucleus, as in the shell model of nuclear physics and nuclear chemistry.

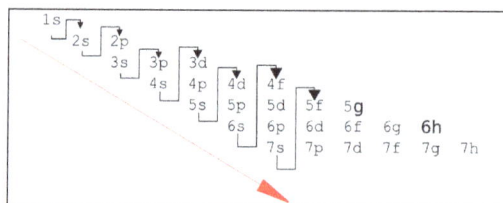

The approximate order of filling of atomic orbitals, following the arrows from 1s to 7p. (After 7p the order includes orbitals outside the range of the diagram, starting with 8s.)

Periodic Table

The form of the periodic table is closely related to the electron configuration of the atoms of the elements. For example, all the elements of group 2 have an electron configuration of [E] ns^2 (where [E] is an inert gas configuration), and have notable similarities in their chemical properties. In general, the periodicity of the periodic table in terms of periodic table blocks is clearly due to the number of electrons (2, 6, 10, 14...) needed to fill s, p, d, and f subshells.

The outermost electron shell is often referred to as the "valence shell" and (to a first approximation) determines the chemical properties. It should be remembered that the similarities in the chemical properties were remarked on more than a century before the idea of electron configuration. It is not clear how far Madelung's rule explains (rather than simply describes) the periodic table, although some properties (such as the common +2 oxidation state in the first row of the transition metals) would obviously be different with a different order of orbital filling.

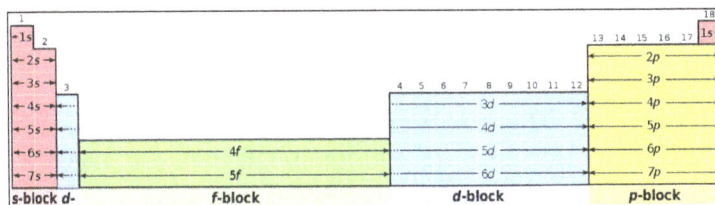

Electron configuration table.

Shortcomings of the Aufbau Principle

The Aufbau principle rests on a fundamental postulate that the order of orbital energies is fixed, both for a given element and between different elements; in both cases this is only approximately true. It considers atomic orbitals as "boxes" of fixed energy into which can be placed two electrons and no more. However, the energy of an electron "in" an atomic orbital depends on the energies of all the other electrons of the atom (or ion, or molecule, etc.). There are no "one-electron solutions" for systems of more than one electron, only a set of many-electron solutions that cannot be calculated exactly (although there are mathematical approximations available, such as the Hartree–Fock method).

The fact that the Aufbau principle is based on an approximation can be seen from the fact that there is an almost-fixed filling order at all, that, within a given shell, the s-orbital is always filled before the p-orbitals. In a hydrogen-like atom, which only has one electron, the s-orbital and the p-orbitals of the same shell have exactly the same energy, to a very good approximation in the absence of external electromagnetic fields. (However, in a real hydrogen atom, the energy levels are slightly split by the magnetic field of the nucleus, and by the quantum electrodynamic effects of the Lamb shift.)

Ionization of the Transition Metals

The naïve application of the Aufbau principle leads to a well-known paradox (or apparent paradox) in the basic chemistry of the transition metals. Potassium and calcium appear in the periodic table before the transition metals, and have electron configurations [Ar] $4s^1$ and [Ar] $4s^2$ respectively, i.e. the 4s-orbital is filled before the 3d-orbital. This is in line with Madelung's rule, as the 4s-orbital has $n+l = 4$ ($n= 4, l= 0$) while the 3d-orbital has $n+l = 5$ ($n = 3$, $l = 2$). After calcium, most neutral atoms in the first series of transition metals (Sc-Zn) have configurations with two 4s electrons, but there are two exceptions. Chromium and copper have electron configurations [Ar] $3d^5 4s^1$ and [Ar] $3d^{10} 4s^1$ respectively, i.e. one electron has passed from the 4s-orbital to a 3d-orbital to generate a half-filled or filled subshell. In this case, the usual explanation is that "half-filled or completely filled subshells are particularly stable arrangements of electrons".

The apparent paradox arises when electrons are removed from the transition metal atoms to form ions. The first electrons to be ionized come not from the 3d-orbital, as one would expect if it were "higher in energy", but from the 4s-orbital. This interchange of electrons between 4s and 3d is found for all atoms of the first series of transition metals. The configurations of the neutral atoms (K, Ca, Sc, Ti, V, Cr) usually follow the order 1s, 2s, 2p, 3s, 3p, 4s, 3d, ...; however the successive stages of ionization of a given atom (such as Fe^{4+}, Fe^{3+}, Fe^{2+}, Fe^+, Fe) usually follow the order 1s, 2s, 2p, 3s, 3p, 3d, 4s, ...

This phenomenon is only paradoxical if it is assumed that the energy order of atomic orbitals is fixed and unaffected by the nuclear charge or by the presence of electrons in other orbitals. If that were the case, the 3d-orbital would have the same energy as the 3p-orbital, as it does in hydrogen, yet it clearly doesn't. There is no special reason why the Fe^{2+} ion should have the same electron configuration as the chromium atom, given that iron has two more protons in its nucleus than chromium, and that the chemistry of the two species is very different. Melrose and Eric Scerri have analyzed the changes of orbital energy with orbital occupations in terms of the two-electron repulsion integrals of the Hartree-Fock method of atomic structure calculation. More recently Scerri has argued that contrary to what is stated in the vast majority of sources, 3d orbitals rather than 4s are in fact preferentially occupied.

Similar ion-like $3d^x 4s^0$ configurations occur in transition metal complexes as described by the simple crystal field theory, even if the metal has oxidation state 0. For example, chromium hexacarbonyl can be described as a chromium atom (not ion) surrounded by six carbon monoxide ligands. The electron configuration of the central chromium atom is described as $3d^6$ with the six electrons filling the three lower-energy d orbitals between the ligands. The other two d orbitals are at higher energy due to the crystal field of the ligands. This picture is consistent with the experimental fact that the complex is diamagnetic, meaning that it has no unpaired electrons. However, in a more accurate description using molecular orbital theory, the d-like orbitals occupied by the six electrons are no longer identical with the d orbitals of the free atom.

Other Exceptions to Madelung's Rule

There are several more exceptions to Madelung's rule among the heavier elements, and as atomic number increases it becomes more and more difficult to find simple explanations such as the stability of half-filled subshells. It is possible to predict most of the exceptions by Hartree–Fock calculations, which are an approximate method for taking account of the effect of the other electrons on orbital energies. For the heavier elements, it is also necessary to take account of the effects of Special Relativity on the energies of the atomic orbitals, as the inner-shell electrons are moving at speeds approaching the speed of light. In general, these relativistic effects tend to decrease the energy of the s-orbitals in relation to the other atomic orbitals. For example, in the transition metals, the 4s orbital is of a higher energy than the 3d orbitals; and in the lanthanides, the 6s is higher than the 4f and 5d.

Electron Configuration in Molecules

In molecules, the situation becomes more complex, as each molecule has a different orbital structure. The molecular orbitals are labelled according to their symmetry, rather than the atomic orbital labels used for atoms and monatomic ions: hence, the electron configuration of the dioxygen molecule, O_2,, is written $1\sigma_g^2\,1\sigma_u^2\,2\sigma_g^2\,2\sigma_u^2\,3\sigma_g^2\,1\pi_u^4\,1\pi_g^2$, or equivalently $1\sigma_g^2\,1\sigma_u^2\,2\sigma_g^2\,2\sigma_u^2\,1\pi_u^4\,3\sigma_g^2\,1\pi_g^2$. The term $1\pi_g^2$ represents the two electrons in the two degenerate ð*-orbitals (antibonding). From Hund's rules, these electrons have parallel spins in the ground state, and so dioxygen has a net magnetic moment (it is paramagnetic). The explanation of the paramagnetism of dioxygen was a major success for molecular orbital theory.

The electronic configuration of polyatomic molecules can change without absorption or emission of a photon through vibronic couplings.

Electron Configuration in Solids

In a solid, the electron states become very numerous. They cease to be discrete, and effectively blend into continuous ranges of possible states (an electron band). The notion of electron configuration ceases to be relevant, and yields to band theory.

Applications

The most widespread application of electron configurations is in the rationalization of chemical properties, in both inorganic and organic chemistry. In effect, electron configurations, along with some simplified form of molecular orbital theory, have become the modern equivalent of the valence concept, describing the number and type of chemical bonds that an atom can be expected to form.

This approach is taken further in computational chemistry, which typically attempts to make quantitative estimates of chemical properties. For many years, most such calculations relied upon the "linear combination of atomic orbitals" (LCAO) approximation, using an ever-larger and more complex basis set of atomic orbitals as the starting point. The last step in such a calculation is the assignment of electrons among the molecular orbitals according to the Aufbau principle. Not all methods in calculational chemistry rely on electron configuration: density functional theory (DFT) is an important example of a method that discards the model.

For atoms or molecules with more than one electron, the motion of electrons are correlated and such a picture is no longer exact. A very large number of electronic configurations are needed to exactly describe any multi-electron system, and no energy can be associated with one single configuration. However, the electronic wave function is usually dominated by a very small number of configurations and therefore the notion of electronic configuration remains essential for multi-electron systems.

A fundamental application of electron configurations is in the interpretation of atomic spectra. In this case, it is necessary to supplement the electron configuration with one or more term symbols, which describe the different energy levels available to an atom. Term symbols can be calculated for any electron configuration, not just the ground-state configuration listed in tables, although not all the energy levels are observed in practice. It is through the analysis of atomic spectra that the ground-state electron configurations of the elements were experimentally determined.

References

- 5.11%3A-Quantum-Mechanical-Atomic-Model,   Book%3A-Introductory-Chemistry-(CK-12)/05%3A-Electrons-in-Atoms, Introductory-Chemistry, Bookshelves: libretexts.org, Retrieved 24 March, 2019

- T. Engel and P. Reid, Physical Chemistry (Pearson Benjamin-Cummings, 2006) ISBN 080533842X, pp. 477–47

- Atomorbs, properties, atoms: chemguide.co.uk, Retrieved 21 March, 2019

- Bohr, Niels (1923). "Über die Anwendung der Quantumtheorie auf den Atombau. I". Zeitschrift für Physik. 13 (1): 117. Bibcode:1923zphy...13..117B. Doi:10.1007/BF01328209.

INDEX

www.ingramcontent.com/pod-product-compliance
Lightning Source LLC
Chambersburg PA
CBHW082035190326
41458CB00010B/3373